I0482832

A Biogeographical Investigation
of the Sierra de Tuxtla in Veracruz, Mexico

by

Robert F. Andrle

1964

Copyright ©2014 Robert F. Andrle
All rights reserved

ISBN: 1496052498
ISBN-13: 978-1496052490

Acknowledgments

To the many persons who aided in this study I extend my thanks, especially to Dr. Robert C. West of the Department of Geography and Anthropology at Louisiana State University for recommendations and critical reading of the manuscript; to Dr. George H. Lowery, Jr., Director of the Museum of Natural Science at the same university, for suggestions and collecting equipment; and to Fred T. Hall, Director of the Museum of Science in Buffalo, New York, for equipment and a leave of absence from the museum staff. I am indebted to officials of several Mexican government agencies for their cooperation, and especially thank Prof. Enrique Beltrán, Subsecretary of Forest Resources and Game, and the late Ing. Luis Macías Arellano, Director General of the Game Department, for their assistance and for granting collecting permits. My sincere thanks go to Dr. Velva E. Rudd, Associate Curator at the U. S. National Museum, for determining my plant collections. Dr. Harold H. Axtell, Curator of Biology at the Buffalo Museum of Science, an excellent field companion, contributed to my work in 1960. I am also indebted to many residents of the Sierra, particularly Prof. Carlos F. Ramírez, Dr. Raphael Gallego de la Cruz and John Lind. Ing. Roberto Gutiérrez Gil of Petroleos Mexicanos, Prof. Othon Arroniz of the University of Veracruz, Dr. Allan R. Phillips, Dr. Robert W. Dickerman, Gary N. Ross, Ruth A. Sparrow, Robert D. Coggeshall and Mrs. Frederick K. Wykes deserve thanks for their help. The encouragement and assistance of my wife, Patricia, was invaluable. Major support for the study was afforded by the National Research Council of the National Academy of Sciences under Subcontract Number 54.

Contents

List of Tables

List of Figures

List of Plates

Abstract

The Sierra de Tuxtla, a small, isolated volcanic mountain area near the Gulf coast of southern Veracruz, Mexico, affords excellent conditions for biogeographical investigations. The present study describes the manner in which physical and human factors have influenced native vegetation, nontransient avifauna and larger forest mammals in this tropical region. Physical factors considered are geology, land configuration, climate, drainage and soils; human factors examined include population, settlement and forms of land use.

Following the Sierra's permanent emergence from marine inundation in the Tertiary, volcanism and erosion gradually transformed the land surface into its present configuration of high-walled craters, steep cones, ridges, valleys and undulating uplands. Most Sierra soils have been derived from basic volcanic materials, and some were formed from sedimentary rocks. The soils' variable depth and composition reflect this complex surface configuration and cause a marked variability in vegetation structure.

High annual rainfall and generally warm temperatures interact with soils to support a rain forest subformation chiefly on the Gulf side of the range. Poorer soils and lower precipitation on southern slopes support forests of gum, pine, oak and semideciduous tropical broadleaf trees. The Gulf of Mexico is a major source of moisture affecting vegetation, and orographic rainfall, especially during northers and easterly waves, is a significant influence.

The aspect of a forested region in a tropical environment was largely maintained through the Sierra's geological history. During over two millennia of human occupance vegetation has been destroyed and modified in more than half the Sierra. Numerous archeological sites reveal the large extent of human settlement, indicating a fairly dense preconquest population mainly in the drier southern slopes and valleys. Here man has created a cultural landscape of fields and planted trees intermixed with forest remnants and thickets. Recent rapid human population increase has been accompanied by expansion of subsistence and plantation agriculture and grazing and removal of natural vegetation. Thus there is now less forest regeneration than when population was less and shifting cultivation more widely practiced.

The proportions of the avian and mammalian faunas having

northern and southern affinities are indicative of the transitional nature of this part of the biota. Pre-human bird and mammal populations were adapted to forest environment and little affected by natural phenomena. Minor endemism developed in birds and several other groups, apparently stemming from isolation since the Pleistocene. Areal distribution of forest and nonforest forms has been radically changed by elimination of forest habitat. This and subsistence hunting have also caused population reductions of some species, but an increase has occurred of those characteristic of nonforest habitats. Avian and mammalian altitudinal distribution is more closely related to vegetation types than to climatic factors or elevation.

The Sierra has no major industrial raw materials, so agriculture continues to be the principal occupation, with commercial plantations providing export products. Although attempts are being made to improve agriculture and conserve forests, immediate and effective measures are required to control land use and enforce game laws. Varied natural habitats and a rich fauna provide excellent bases for a national park or wildlife refuge. Such an area would conserve forests, soils, water and wildlife, and afford a place for scientific study and recreation.

A working man-land relationship must be established if the Sierra's natural resources are to be used wisely, its agricultural productivity increased and natural features preserved. The area should be considered a distinct physical and biological unit and its potential evaluated in terms of the foregoing factors. A concerted effort to raise the educational level of many Sierra inhabitants should also be undertaken in conjunction with a long-range program of proper land use and conservation.

Introduction

The Sierra de Tuxtla is a volcanic highland lying along the Gulf of Mexico about 80 kilometers southeast of Veracruz city near the northern end of the Isthmus of Tehuantepec (Figure 1). The range trends northwest-southeast and is approximately 90 by 50 kilometers in maximum areal dimensions. Four large volcanoes, the highest 1660 meters in elevation, comprise the major parts of the highland. There are several lower peaks and many volcanic cones and foothill ridges. A central basin contains Lago Catemaco, the third largest lake in Mexico. Extensive lowlands drained by the Coatzacoalcos and Papaloapan river systems border the mountains on the landward side. During prior studies in other parts of Mexico I made a brief visit to the Sierra de Tuxtla. My return in 1960 served to strengthen the idea which occurred on my first trip, that it is a region well suited for biogeographical study because of its coastal location, physical isolation, and rich flora and fauna; equally important are the various cultural influences that have caused major changes in its environment. With principal support from the National Academy of Sciences I was able to work in the Sierra from February to December 1962.

The mountain range as a unit has no official name. The northwestern section is simply called "Los Tuxtlas" after the former canton encompassing this area; the southeastern massif (Volcán Santa Marta) lies largely in the former canton of Acayucan and has been called the Sierra San Juan or Sierra de Acayucan. Because these mountains form a single topographic uplift, the name Sierra de Tuxtla will be used to refer to the entire range. Lago Catemaco separates the range into essentially two portions which, for purposes of this study, I have designated as the Volcán San Martín Tuxtla massif and the Volcán

1

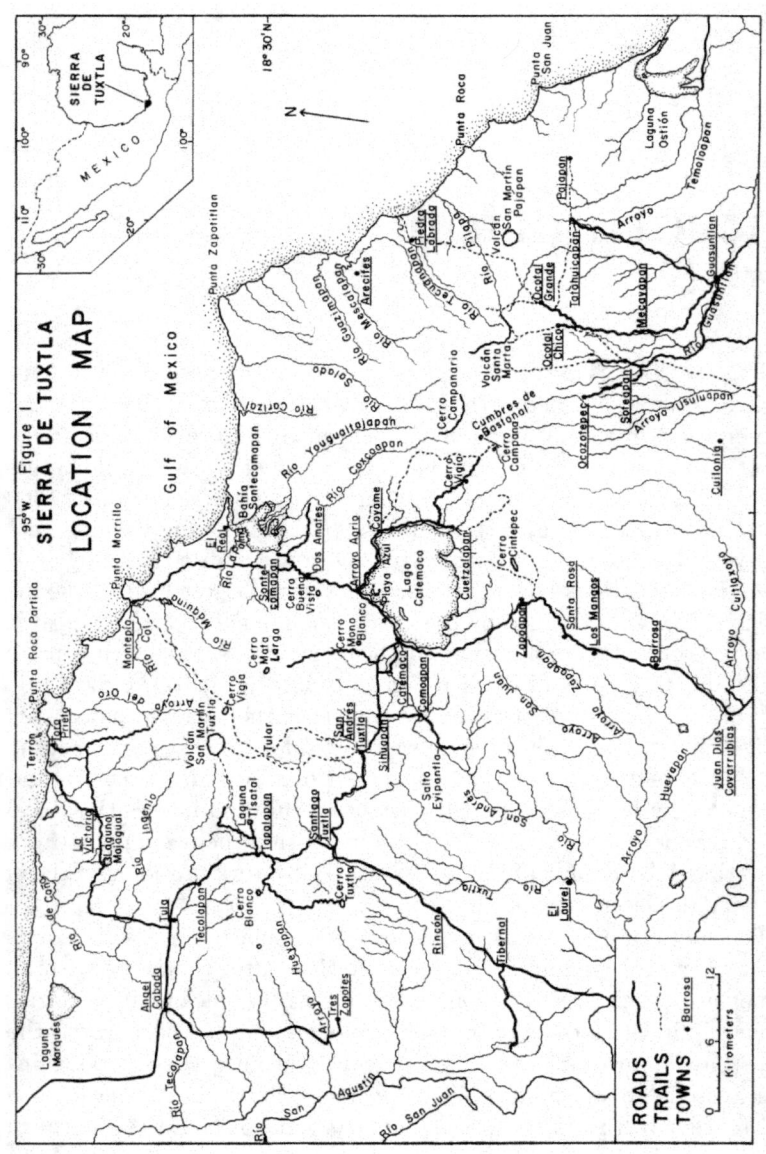

SIERRA DE TUXTLA
LOCATION MAP

Figure 1

Gulf of Mexico

ROADS
TRAILS
TOWNS • Barrosa

0   6   12
Kilometers

2

Santa Marta massif.

Purposes and scope. The purpose of this investigation is to show how physical and human factors have influenced the native vegetation, nontransient avifauna and larger tropical forest mammals of the region. The physical factors considered include geology, climate, land configuration, drainage and soils; the human factors emphasized are population densities, settlement types and forms of land use. Volcanism and forest destruction by man are here considered to be two of the most significant elements that have induced floral and faunal changes within the Sierra de Tuxtla.

The wildlife in these tropical mountains forms a conspicuous and important environmental component. Only nontransient birds and certain of the larger forest dwelling mammals are treated in this study not only because of my interest in them, but primarily because they are the prominent elements of the fauna that are radically affected by man as an agent of destruction and as a consumer.

The conservation of renewable natural resources in Mexico is important to the future of the country. Intimately linked with the proper use of natural resources is the need for the preservation of undisturbed natural areas. The Sierra de Tuxtla was recommended by Shelford (1941:109) and Leopold (1959:93) as an excellent place in which to establish a national park, a view with which I concur. Enrique Beltrán, Subsecretario de Recursos Forestales y de Caza, who has long worked for such objectives in Mexico, is especially interested in this region. Appendix A contains my estimate on the feasibility of establishing such a park and a map outlining the areas which I think are most suitable.

Previous research. Scientific studies in the Sierra de Tuxtla have been brief, areally restricted, and confined in each instance to one or two disciplines. From Sclater (1857) came the first reports of collections of birds from San Andrés Tuxtla and Sontecomapan. Accounts of the two volcanic disturbances of historic record were given by García (1835), Zérega (1870) and Mociño (1870, 1874-76). Kerber (1882) reported on an old village site near Montepío on the Gulf coast. In 1894 Nelson and Goldman (1951) observed and collected mammals and birds for about three weeks at Catemaco, San Andrés Tuxtla and on Volcán San Martín Tuxtla (formerly called *Titépetl*). Seler-Sachs (1922) described some antiquities from the region. Friedlaender (1923) wrote an informative paper on the geology and surface configuration; his map of the Sierra is remarkably accurate from an areal aspect. The Tulane University Expedition to Middle America (Blom and La Farge, 1926, 1927) visited the region, making brief ethnological and archeological studies. Schieferdecker and Tschopp (1922), Tschopp (1926, 1931), Schumacher (1929) and Staehelin (1935) treated geologic aspects in connection with the search for minerals. Foster (1940, 1942) wrote accounts of the Popoluca Indians and their country on the southern slopes of the Sierra. Using Tres Zapotes as a base in the lowlands on the southwest, Wetmore (1943) and Carriker made ornithological collections in the late winter and spring of 1939 and 1941, chiefly on Cerro Tuxtla and Volcán San Martín Tuxtla. Accounts of archeological investigations in several localities were given by Valenzuela (1939, 1945). Ríos Macbeth (1952) made a geological study in part of the range. In 1954 Edwards and Tashian (1959) studied the bird life on the northeastern side of Lago Catemaco. Several ornithologists, entomologists and herpetologists have published their investigations in various parts of the northwestern and southern sections of the Sierra (Davis, 1952; Goodnight and Goodnight, 1954; Amadon and Eckelberry, 1955; Firschein, 1950; Firschein and Smith, 1956), and Phillips and Dickerman (1962, 1963) have afforded additional ornithological information from their observations during those years.

My first contacts with the region were in 1951 and 1952 when I made brief reconnaissances. I spent about a month there in March and April 1960, studying certain avifaunal aspects. This investigation is the first biogeographical study undertaken in the Sierra and is intended to present the range (about 4500 square kilometers) as a composite unit possessing distinct physical, biological and cultural characteristics.

Methods. In February 1962 I established a main base of operations at Playa Azul on the western shore of Lago Catemaco (Figure 1). This central location facilitated extension of investigations in many directions into the mountain range. In general I directed my studies first to the Volcán San Martín Tuxtla massif, second to the Lago Catemaco-Bahía Sontecomapan area and third to the Volcán San Marta massif. At times I was compelled to vary the sequence, weather and availability of transportation being the principal governing factors. I established

4

boundaries for the study at the coastline on the Gulf side and about 100 meters elevation on the landward sides; this enabled me to include the steep slopes and headlands which are an essential part of the Sierra facing the Gulf of Mexico. On the few existing roads and wider trails I employed two and four-wheel drive vehicles. Elsewhere I travelled on foot, by horse and mule and by power boat and canoe. One aerial reconnaissance was made. Time away from main base varied from several hours to a maximum of ten days, depending on the sections scheduled to be covered. I set up subsidiary bases in the more remote areas when it was necessary to work for several days or more in one locality.

The adequate coverage and effective sampling of an area encompassing about 4500 square kilometers, much of which is difficult of access, necessitated relatively simple methods and a minimum of equipment. My essential equipment consisted of pen and notebook, compass, barometer-altimeter, binoculars, camera, shotgun and specimen bags. Visual and auditory observations combined with correlative analysis were the principal methods used throughout the study. Whenever possible, altitudes of mountains, lakes, settlements and other features were determined carefully with the barometer-altimeter. Many of these points were rechecked in order to secure as accurate data as possible. At every opportunity I obtained information from local people in villages and on roads and trails. Bird and plant specimens were selectively secured and photographs taken whenever appropriate to the objectives of the investigation. At the main base I employed a maximum-minimum thermometer and a rain gauge. A previously drawn base map was modified as necessary and used with overlays on which I recorded information obtained in the field pertaining to the various phases of the study.

Chapter 1

PHYSICAL ENVIRONMENT

1.1 Geologic Factors

The geologic history of the Sierra de Tuxtla is only generally known. Tschopp (1931) referred to the basement ridge (probably Paleozoic sedimentary and crystalline rocks) of these mountains as existing in the early Mesozoic Era, and Staehelin (1935) indicated that it must have acted as a barrier against the advance of the sea which inundated part of southern Mexico (the Isthmian Embayment and Veracruz Basin) in the Jurassic Period upon the formation of the Mexican geosyncline. There is no geological evidence that the Sierra was ever connected to other mountain ranges by a high ridge, although it is possible that there was a basement highland to the Chiapas mountains in late Mesozoic or early Tertiary. Murray (1961:135) considered the Tuxtla uplift, together with others such as the Teziutlan and North Yucatan, as probably high areas of basement rock in the Mesozoic-Cenozoic geosyncline, and as part of the arc-shaped "Tamaulipas-Yucatan archipelago" of highs which have influenced the topography and configuration of the coastal province. He indicated that these highs possibly represent rejuvenation by isostatic adjustment to the accumulations of thick sedimentaries in adjacent portions of the geosyncline. The syncline tended to sink and the region was subjected to a series of marine transgressions involving much of the Gulf Coastal Plain until the latter's gradual general emergence during the late Tertiary Period (Miocene-Pliocene). Thus a thick series of Cretaceous and Tertiary deposits of blue clays and shales, tuffs, sandstones, limestones and conglomerates was laid down during these times over the pre-Cretaceous basement of the geosyncline, and upon emergence was partially eroded. The Oligocene and Miocene sedimentary rocks which have been found exposed in the Sierra are chiefly arenaceous-argillaceous-calcareous materials (Murray, 1961:153). Although the geology of the Sierra has not been studied thoroughly, Figure 2 shows the locations of volcanic cones, lava flows and the exposed Tertiary sedimentaries which have been found within the cover of volcanic materials.

The dioritic rocks found by Friedlaender on the Arroyo del Oro and by Tschopp near Tecolapan suggested to Schieferdecker and Tschopp (1922) that this volcanic region rests on a dioritic laccolith of early Miocene or Oligocene age which has lifted and in places folded the Tertiary beds, and from which the volcanic extrusives have emerged. These extrusives comprise most of the Sierra and consist of older Pliocene material of an acid andesitic nature (exposed at various places, for example, Punta Roca and Arroyo Pilapa) upon which were superimposed basic basalts of Quaternary age. Schuchert (1935:374), in describing the Atlantic coastal plain, said that it is composed of nonfolded sedimentaries and "a strip of very young basalt and ash near the coast which rises into the volcano of San Martin (5700 feet)." The depth of this cover is variable. Exposures in road cuts and gravel pits show layers three or four meters to over 50 meters thick. It is probable that the volcanic materials are considerably deeper than this in places. In some localities, particularly near the coast, these basalts are overlain by Pleistocene quartz sands and gravels. Pliocene-Pleistocene deposits border the inland sides of the Sierra and rest uncomfortably on older Tertiary sequences. Staehelin said that no Cretaceous fold or any deposits older than Oligocene are known from the plunging parts of the "San Andres Tuxtla" [Volcán San Martín] massif. This would indicate that at least part of the Sierra was emergent during the Mesozoic and early Tertiary.

It can be concluded that the present Sierra de Tuxtla volcanic uplift composed of older andesitic and more recent basalt flows and volcanic

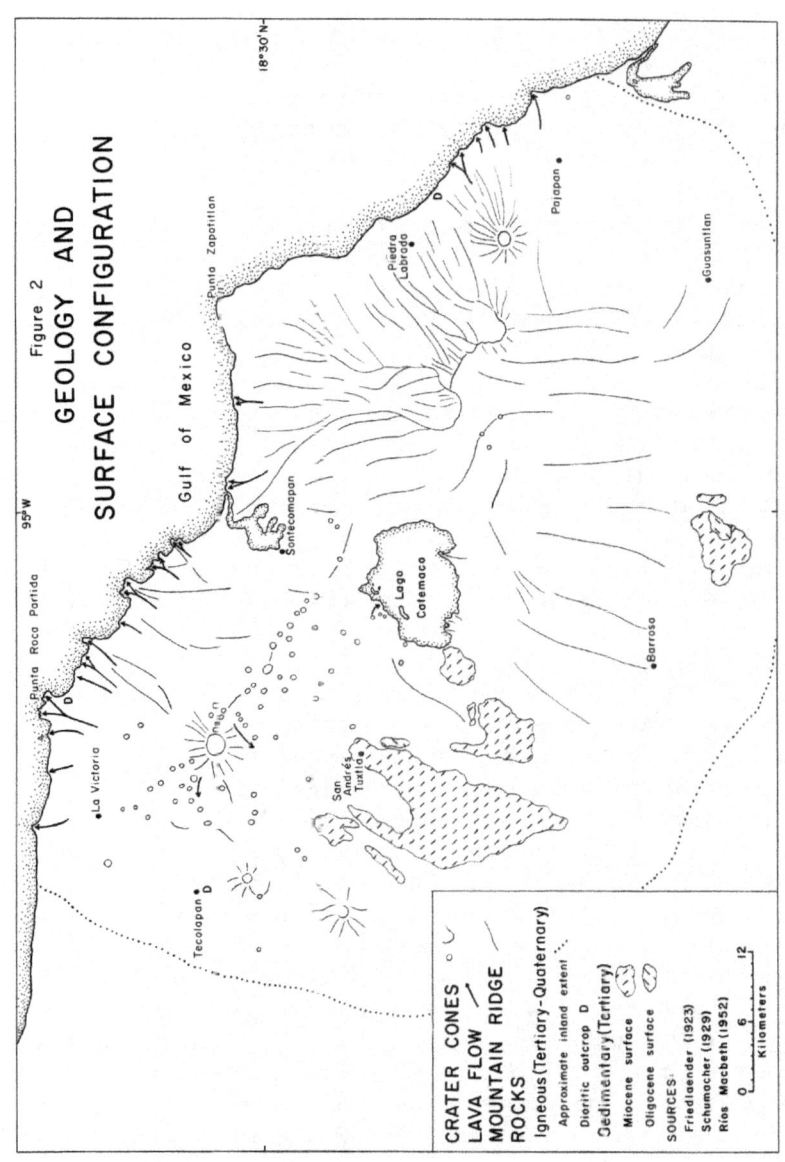

Figure 2

GEOLOGY AND
SURFACE CONFIGURATION

CRATER CONES
LAVA FLOW
MOUNTAIN RIDGE
ROCKS
Igneous (Tertiary-Quaternary)
Approximate inland extent
Dioritic outcrop D
Sedimentary (Tertiary)
Miocene surface
Oligocene surface
SOURCES:
Friedlaender (1923)
Schumacher (1929)
Rios Macbeth (1952)

0   6   12
Kilometers

9

plugs is of late Pliocene age and was formed after the last marine inundation. The volcanic activity in this range in the late Cenozoic was paralleled in some of the other mountainous areas of the country to the north and west, although igneous activity in these areas was more or less periodic throughout Cenozoic time at least as early as the Oligocene Epoch.

Seven principal eruption centers or zones are evident in the Sierra de Tuxtla, although other older ones have likely been buried. Those recognizable are: Cerro Tuxtla (825 m. above sea level), Cerro Blanco (720 m.), Volcán San Martín Tuxtla (1660 m.), the Lago Catemaco Basin (including Cerros Mono Blanco, Las Ánimas and Cintepec (890 m.), Cerro Campanario (1180 m.), Volcán Santa Marta (1550 m.), and Volcán San Martín Pajapan (1270 m.). Numerous subsidiary lava, ash and cinder cones exist, particularly in the vicinity of Volcán San Martín Tuxtla and to the southeast of this volcano. Fine-grained olivine basalt rocks are the dominant type in every eruption center. Friedlaender (1923) tentatively established two lines of eruption centers north of Santiago Tuxtla and San Andrés Tuxtla. From aerial photographs a major line also appears to have been toward the southeast (Figure 2). Lago Catemaco (Plate I) was considered by Friedlaender to be a caldera. Aerial photographs show that the lake contains two exposed rims of craters similar to that of Cerro Mono Blanco, and appears in perspective from the south to be no more than a stream and spring filled section of the range whose southern and western borders have been effectively blocked by volcanic cones and debris. The outlet to Lago Catemaco and some of the streams that flow from the northwest section of the massif form waterfalls over basalt cliffs. To the northward lava flows have formed ridges, some fronting on the Gulf shore in steep headlands (Plate II). Large basalt blocks are visible in many places, particularly in stream bed outcrops and where there is no forest cover such as in the lava flow southeast of Volcán San Martín Tuxtla. There are several steep, bare ash slopes high on Volcán San Martín Tuxtla. Layers of ash, lapilli and cinders are exposed in road cuts and gravel pits, particularly in the Catemaco section. Basalt bombs, low lava projections and pumice also occur in many places, the latter two particularly near the coast (Plate III). Also scattered along the Gulf shore are asphalt cakes of various sizes apparently from petroliferous Tertiary sandstone beds.

Plate I. Lago Catemaco

Colonia Adalberto Tejeda, 520 m. a.s.l. This is a view westward along
the north side of the lake showing some of the rain forest above Coyame.
Isla Agaltepec, at the left, is part of the crater rim of a volcanic cone.
At the right, the buildings of Finca Rociela are visible with Cerro Mono
Blanco the middle cone in the distance.

Plate II. Volcanic Headland

Aerial photograph of Punta Morrillo on the Gulf of Mexico. Cliffed headlands sometimes over 30 meters in height are a topographic feature of the Sierra shoreline. The ends of these lava flows are usually covered with a low, dense littoral vegetation. Between the headlands are sand beaches, such as the one at right where the Ríos Máquina and Col enter the Gulf at Montepío.

Plate III. Lava Flow at Gulf of Mexico

Sea level, 5.5 km. west Punta Roca Partida. Low lava projections occur at intervals along the coast. Much of this lava has formed into the pillow type shown here.

The large composite volcanoes of Cerro Campanario, Volcán Santa Marta and Volcán San Martín Pajapan have no historical record of eruptions, and I was not able to perceive any evidence of recent volcanic disturbance. They all have large deep craters and steep lava ridges radiating mostly toward the Gulf coast. The first two mentioned have their crater walls broken or absent on the north because the major eruptions occurred on their north flanks with lava flows breaking out in that direction. The broader ridges to the south of these volcanoes are composed of older, much weathered, basalt lava. Small eruption cones lie in the craters of Volcán Santa Marta and Volcán San Martín Tuxtla. The latter is known to have erupted on October 15, 1664, and again beginning on March 2, 1793. Medel y Alvarado (1963: [1] 29-30) speaks of "the violent eruption of Titépetl at the termination of the first third of the 16th century."

Little has been recorded concerning the 1664 disturbance. Mociño Suarez de Figueroa writes (Medel y Alvarado, 1963: [1] 80) that in this eruption "the materials did not pass three leagues." Friedlaender mentioned that it was described as an ash eruption, but he believed a lava stream poured out to the north. He attributed an older block lava that he saw to the northeast to this eruption. The most detailed account of the 1793 activity was given by Don José Mariano Mociño de Figueroa (1870), botanist to the Royal Expedition of New Spain. He tells of violent explosions accompanied by heavy ash fall taking place at intervals from March into September. In October the two small cones in the crater were still active, although considerably less so than during the early part of the eruption. In 1829 José Aurelio García observed fumarolic activity in the crater. The lava of the 1793 eruption flowed out of the crater to the northeast and northwest. Block lava is exposed now on the crater floor, and in the forest on the crater edges. There are vertical-walled lava channels in this volcano's crater, and tunnels on Cerro Mono Blanco and at Laguna Encantada near San Andrés Tuxtla.

1.2   Land Configuration and Drainage

The Sierra is a discrete mountain mass dominated by four large volcanoes. They culminate an uplift which slopes gradually, but irreg-

14

ularly, to the Gulf Coastal Plain on all sides except the seaward where in several places long and often steep-sided ridges extend to the Gulf shore. Figure 3 shows the general elevation patterns in the mountain range. Cerro Tuxtla and Cerro Blanco, two larger outlying volcanoes on the southwestern border, interrupt the descent to the lowlands in this direction. The northwestern massif, centered on Volcán San Martín Tuxtla, exhibits a surface with many medium to small sized, cratered cones and rounded hills (Plate IV); these are interspersed among the above mentioned ridges on the north and with undulating terrain and variously aligned valleys and ridges on the southern side. Some of these cones, as well as the larger volcanoes, have rather deeply dissected slopes and their ravines exhibit a radial erosion pattern. There are at least ten water-filled craters such as Lagunas Tisatal (Plate V) and Majagual. Although Cerros Tuxtla and Blanco show only remnants of craters (partially walled), San Martín Tuxtla possesses a large, deep crater, lower walled on the north with two former lava flow gaps. It is oval in shape, about 1.5 kilometers maximum width and 200 meters in depth (Plate VI). A distinct line of large cones, some cratered, begins with Cerro Vigía (ca.1400 m.) and extends southeastward toward Lago Catemaco. Cerros Mata Larga and Buena Vista (700 m.) are also prominent on this line.

On the west and southwest sides of the large basin of Lago Catemaco are several cones and crater lakes. Islas Agaltepec and Tenaspi in Lago Catemaco are the exposed rims of craters. Rounded hills and small valleys occur to the south of the lake with the broad, elongated uplift of Cerro Cintepec on the southeast. South of this mountain long rounded ridges and narrow valleys descend to the lowlands. The north side of the Catemaco basin is bordered by a ridge of rounded hills, the extension of the line of cones leading southeast from Volcán San Martín Tuxtla. The large, mangrove-fringed Bahía Sontecomapan is an irregular shaped bay of the Gulf north of Lago Catemaco (Plate VII). Its small adjoining lowland on the west is bordered by ridges and low hills, and a long, steep-walled lava ridge from the Volcán Santa Marta massif borders it closely on the east. On the east side of Lago Catemaco the terrain rises abruptly into the southeastern part of the Sierra, broken only on the southeast by the deep gorge of the Río Cuetzalapan.

The north slopes of the three large crater volcanoes are characterized

15

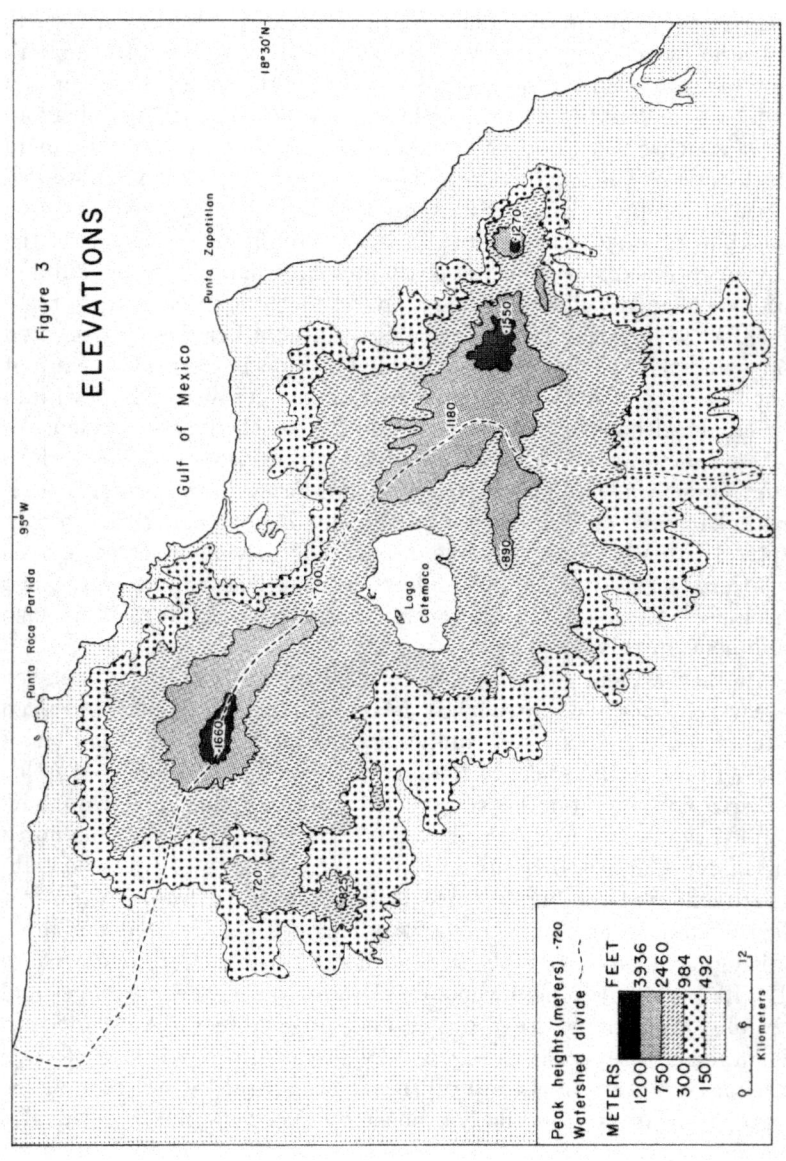

Figure 3

ELEVATIONS

Gulf of Mexico

95°W

18°30'N

Punta Roca Partida

Punta Zapotitlan

Lago
Catemaco

Peak heights (meters) ·720
Watershed divide ----

METERS		FEET
1200		3936
750		2460
300		984
150		492

0   6   12
Kilometers

Plate IV. Surface Configuration

Aerial view west-northwest from Lago Catemaco. This photograph
shows the numerous volcanic cones superimposed on gently undulating
terrain gradually rising toward the highest peaks on the right. Forest
remnants, particularly on the cones, are scattered throughout, and the
edge of the continuous humid forest is visible at the upper right.

17

Plate V. Crater Lake

3 km. east-northeast Tapalapan, 380 m. a.s.l. Laguna Tisatal is one of about ten such lakes in the Sierra. Most are circular and vary from about 60 m. to almost a kilometer in diameter. Many are deep and some have reeds and small areas of marsh vegetation on their borders. These two men are retrieving grebes *(Podiceps dominicus)* which they have shot for food.

Plate VI. Volcano Crater

Volcán San Martín Tuxtla, 1650 m. a.s.l. This is a view toward the northwest showing part of the main crater. The steep, high south wall at left is covered with elfin forest and dense thickets. This crater's walls vary considerably in height as do those of the other large volcanoes. One of the cloud banks that frequently move in from the Gulf is covering the small cones in the distance and will soon envelop the volcano.

by deep ravines and dissected lava ridges radiating toward the Gulf where there occur the widest sections of relatively lower land ranging from sea level to about one-hundred meters in elevation. Volcán San Martín Pajapan's crater is smaller than that of Volcán San Martín Tuxtla, and is less open to the north than are those of Cerro Campanario and Volcán Santa Marta.

The last two volcanoes possess mainly semicircular crater walls on the south. Wide, irregular ridges and in places steeply walled valleys and ravines extend far southward from this massif and gradually merge inland into the plain of the Río Chacalapan, a tributary of the Río Coatzacoalcos. Toward Arroyo Hueyapan and the Río Chacalapan, there are few volcanic cones or hills. To the southwest are several, such as Cerro Campana and Cerro Vigía, on the curving ridge of the Bastonal.

The Sierra de Tuxtla is a well-watered region containing many permanent streams, some intermittent ones, and the previously mentioned crater lakes. The watershed divide between the Papaloapan Basin and streams flowing into the Coatzacoalcos Basin and the Gulf is shown in Figure 3. In the northwest the swift flowing Tecolapan, Tuxtla, Salinas and Máquina rivers and the Arroyo del Oro descend from the Volcán San Martín Tuxtla massif (Figure 1). Most of these streams drop over small basalt cliffs and ledges (Plate VIII). The Río San Andrés (Grande) is the only outlet from Lago Catemaco. It too is permanent and fast flowing and drops over several cliffs to form small waterfalls and the higher one of Eyipantla (ca. 40 meters). Several arroyos, such as the San Juan and Zapoapan, lead southwestward from the Sierra. Some of these are partially dry during certain times of the year. The Río Cuetzalapan is the major stream flowing into Lago Catemaco, coming in from the southeast through a deep valley bordered on the east by the Cumbres de Bastonal. There are also at least two highly mineralized springs (Arroyo Agrio, Coyame) on the north side of the lake with comparatively large volumes of flow.

A number of major permanent streams have their sources in the southeastern massif (Plate IX). Two of these, the Coxcoapan and Yougualtajapan rivers, have long courses leading northwest into the Bahía Sontecomapan. The larger streams such as the Carizal, Salado, Mescalapan, and Tecuanapan are also rapid and, like most of the Gulf-flowing rivers, have their courses determined chiefly by the

20

Plate VII. Bahia Sontecomapan

Aerial view inland toward the south-southwest. This photograph shows how the course of the bay's entrance channel is affected by the volcanic lava ridge projecting toward the right. The Gulf shore is at lower right. Lago Catemaco is partly visible in the upper left with the forested hill ridge connecting the two principal massifs showing dark above the extensive cleared sections south and southwest of the Bahía. The two irregular points projecting from the left into the bay are derived partly from deposition by the Ríos Yougualtajapan and Coxcoapan.

directional trends of the volcanic ridges. Consequently, there are few large tributaries and little stream meandering on the Gulf slope; some meandering occurs in the comparatively narrow, more level areas adjacent to the coast such as in the vicinity of Punta Zapotitlan, Piedra Labrada and the Bahía Sontecomapan. The Gulf-flowing streams enter the sea at a low gradient and their mouths are subject to shifting under the influence of wind, waves and stream flow variation. South and east from this section of the range there is a fairly high density of dendritic streams, some intermittent, emptying into the Río Chacalapan and into the Laguna Ostión. The principal permanent stream is the Río Guasuntlan with others such as the Usuluapan and Temoloapan prominent (Figure 1).

The large streams in the Sierra de Tuxtla have regimes with high volumes during and shortly after periods of heavy rainfall from June to August and from October to December (Comisión del Papaloapan, 1958). In the Volcán San Martín Tuxtla massif there are no permanent streams at the higher elevations above about 1200 meters. In the southeast massif streams occur at elevations nearer the volcano summits. For example, a tributary of the Río Tecuanapan flows from the crater of Volcán Santa Marta at an elevation of about 1300 meters. The southeastern section of the Sierra de Tuxtla experiences more rapid runoff of precipitation, especially on the southern slopes, than the Volcán San Martín Tuxtla massif where deeper and more recent ash layers appear to provide a more porous surface, particularly at the upper elevations. Lago Catemaco (325 m. a.s.l.), as well as the various crater lakes, fluctuates in level from the influence of the rains, the former sometimes rising as much as three feet after heavy precipitation. Reports from local inhabitants that crater lakes, such as Laguna Tisatal, attain a higher level during the dry season possibly indicate a slow downward water movement from rainy season precipitation through subsurface channels to the lakes, which are all below 800 meters elevation.

1.3   Climatic Factors

The Sierra de Tuxtla's position in relation to the Atlantic subtropical high pressure cell and the North American land mass, its proximity to the Gulf of Mexico and its local relief are factors affecting

Plate VIII. Salto

5.5 km. northeast Tapalapan, 490 m. a.s.l. On the slopes of the Sierra
there are falls such as this one on a tributary of the Río Tecolapan;
they drop over outcroppings or cliffs of basalt. This falls is about six
meters high, and others attain various heights up to 40 meters.

Plate IX. Mountain Stream

1.5 km. east-southeast Volcán Santa Marta, 500 m. a.s.l. Swift flowing, cool streams descend from high elevations in the Sierra. Some are intermittent, becoming dry or nearly so during part of the year. Large basalt boulders, as shown here, and volcanic ash and gravels often occur in stream beds.

climatic conditions there. The range is subject to northeast trade winds and occasional easterly waves from the Atlantic high and to northers *(nortes)* from North American polar air masses. It is also affected by the moderating influence of the warm Gulf waters and by wind and air mass movements from inland Mexico. Each of the foregoing interacts, alone or in combination at different times of the year, with the varied topography of the mountains. One of the principal results is an orographic modification which is a distinct characteristic of the climate in the Sierra. The four large volcanoes with their adjacent subsidiary cones and ridges present a northwest-southeast trending front whose crest varies from 360 to over 1600 meters in height and is about 55 kilometers in length; this presents a barrier to winds and air masses from both the Gulf and from the lowlands to the south. Although the climate of these mountains is mainly humid tropical, drier conditions exist in certain sections, especially at the lower elevations and on the inland slopes. There are several meteorological stations in the mountains, but none on the Gulf slopes. There are a few stations in the surrounding low country inland from the uplift. Only two of the mountain stations possess data for 15 or more years.

Wind. The prevailing wind in the region is from the northeast. The land configuration and possibly Lago Catemaco appear to cause the variability in wind direction which is evident locally in the mountains. Frequently south and south-southwest winds occur, especially from late winter into the summer. There is a slight sea breeze at times toward dusk at lower elevations on the Gulf slope, and air drainage downslope was noted at higher elevations on several occasions.

The most notable phenomenon in relation to wind, however, is the conflict which takes place between northerly and southerly components over the Sierra. Griscom (1932:19-20) observed a similar situation in the mountains in Guatemala. Beginning in May, as insolation increases, and continuing into October, the northeast wind exerts its influence on the northern side of the Tuxtlas and meets an inflow of southerly air over the higher elevations. There appear to be daily oscillations in which each component becomes dominant. When the northeast wind is fairly strong and constant, the wind and towering cumulus clouds from inland to the south do not encroach significantly on the Sierra. When the northeast wind decreases, particularly during the night, the

25

south wind may increase and the vertical development of cumulus clouds is sufficient to cause local heavy rain in the mountains. The northeast wind may be accompanied by almost clear skies or on other occasions may be sufficiently moisture laden to cause heavy clouds and showers as air rises over the mountains. On some days in July the northeast wind increases toward evening, possibly the result of high insolation and the resultant lower pressure inland; at such times light to heavy night rainfall takes place. From an examination of the daily weather maps for 1962, published by the Servicio Meteorológico Mexicano, it was clear that this wind conflict, although in part due to local conditions, is also the result of the Sierra's position on the fringes of three pressure systems, a) the Atlantic subtropical high pressure cell (northeast winds), b) the low pressure center over north-central Mexico and c) the low center over the Pacific and the southern part of the Isthmus of Tehuantepec (southerly winds). Their oscillations in position and changes in intensity, particularly from late April to October, cause sufficient pressure change over the Sierra to bring on this wind alternation. During this period the Atlantic high cell moves south and westward with more constant northeast winds, the Pacific low cell more often shifts northward across the isthmus increasing the south wind component, and the low cell over central and northern Mexico expands and occasionally intensifies. The variations in position and intensity of the last named system seem to correlate directly with the strength of the south wind over the Sierra. At other seasons, for example, in February 1962, a pronounced southerly wind occurs when the Atlantic high is more distant and the influence of the two low systems predominates.

Temperature. The average annual temperature in the Sierra, derived from data at the six stations in the mountains (average elevation 290 meters), is about 24.2 degrees Centigrade. This is slightly above the range quoted by Richards (1957:137), (19-24°C.) for areas occupied by tropical rain forest between 200 and 1000 meters elevations. The average mean of all six stations for the coldest month is 20.1 degrees Centigrade, slightly above the mean of 18 degrees which Koppen regarded as the lower temperature limit for his "Tropical Rain Forest Climate."

In examining the average monthly temperatures of all weather sta-

26

tions in the Sierra (San Andrés Tuxtla at 360 meters is the highest above sea level) I found that there is a comparatively low mean annual range, 5.4-8.0 degrees Centigrade. Richards (1957:136) said that "even in the least equable tropical climates it [seasonal variation in temperature] seldom amounts to more than 13 degrees [C.]." It must be remembered, however, that much of the Sierra lies above 360 meters elevation. The lowest temperature recorded at the station (San Andrés Tuxtla) at this altitude was 6.8 degrees Centigrade. Assuming a temperature decrease of 0.5°C. per 100 meters elevation, the minimum at this time at the highest point of the range would have been about zero degrees Centigrade. On February 9 and 10, 1899, there was a severe norther, and the inhabitants of Belem Grande (suburb of San Andrés Tuxtla) reported the top of Volcán San Martín Tuxtla covered with ice (Medel y Alvarado, 1963:[_] 446-447). The highest temperature recorded at the same station was 42.6 degrees Centigrade. It is thus evident that temperature range may be relatively large when changes in elevation and varying exposure are taken into account. The moderating effect on temperature of the Gulf of Mexico is an important factor in the Sierra de Tuxtla. Although freezing has occurred at the upper levels on Volcán San Martín Tuxtla and probably on the other large volcanoes, it is unlikely that it is frequent or prolonged. The warm Gulf water, frequent cloud cover at the higher altitudes, and the extensive forests contribute toward lowering the temperature range. The northers which occur from October to April bring markedly lower temperatures and contribute toward reducing the monthly means during part of the drier period of the year. Counteracting this is the increasing insolation as the dry season progresses.

May and June are usually the warmest months (higher insolation and less cloud cover), January and February, the coldest, but the extreme daily minimum has been most often recorded in February and the extreme maximum in May. Figure 4 shows the monthly progression of temperatures at two stations, San Andrés Tuxtla and Coyame, and Table 1 shows temperature and precipitation data for stations within the mountain range. The maximum and minimum temperatures I recorded at Playa Azul, the main base (340 meters), were 33.8 (April and May) and 17.2 (March) degrees Centigrade respectively from mid-February to mid-December. In this sheltered location in close proximity to Lago

27

TEMPERATURE AND PRECIPITATION

Figure 4

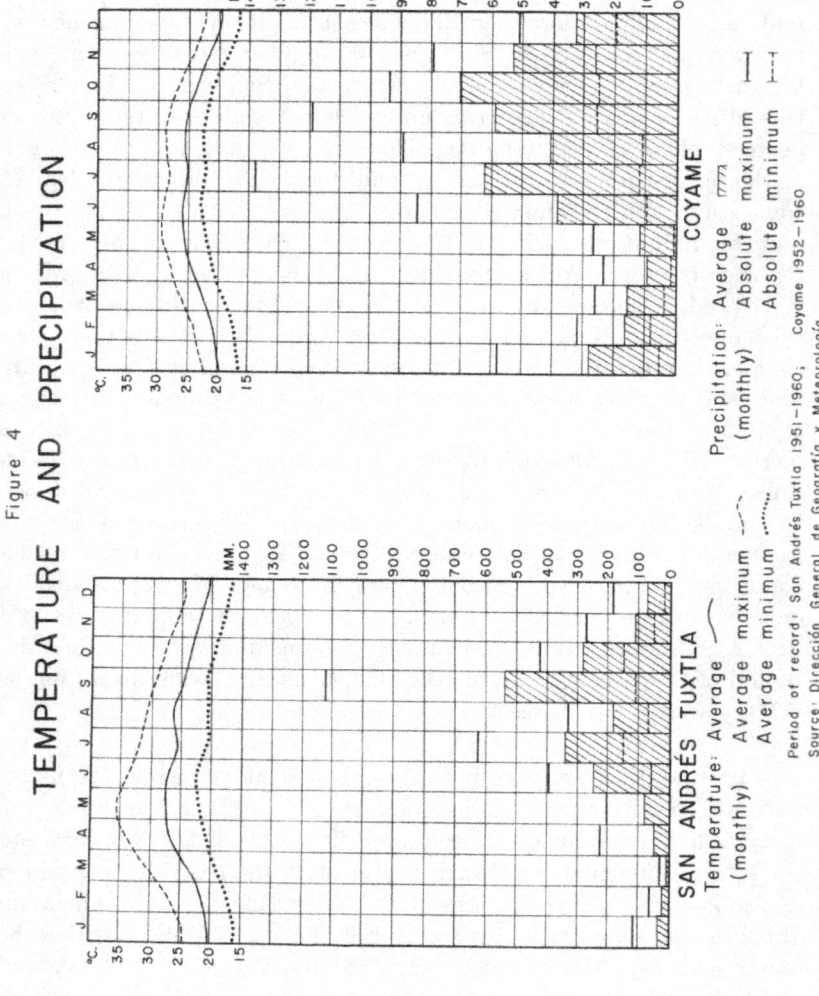

SAN ANDRÉS TUXTLA

Temperature: Average ———
(monthly)   Average maximum ---
            Average minimum ·······

COYAME

Precipitation: Average ▨
(monthly)   Absolute maximum I
            Absolute minimum ⊢—⊣

Period of record: San Andrés Tuxtla 1951—1960;   Coyame 1952—1960

Source: Dirección General de Geografía y Meteorología

28

Catemaco the range between maximum and minimum temperatures as well as the maximum daily range (10.5 degrees Centigrade during present work) are usually less than those at a more exposed station like San Andrés Tuxtla.

TABLE I
SIERRA DE TUXTLA STATION METEOROLOGICAL DATA*

	Length of Record (yrs.)	Elevation (m.)	Highest Maxi-mum (°C.)	Lowest Mini-mum (°C.)	Mean An-nual (°C.)
Santiago Tuxtla	10	210	41.0	8.0	24.4
San Andrés Tuxtla	32	360	42.6	6.8	24.4
Tapalapan	3	290	39.0	8.0	24.4
Coyame	9	340	38.5	10.0	23.4
Catemaco	16	354	42.0	8.0	23.9
Guasuntlan	6	180	40.0	10.0	24.9

	Mean of hottest Month (May or June) (°C)	Mean of coldest Month (Jan or Feb) (°C)	Mean Annual Rainfall (mm.)
Santiago Tuxtla	26.6	19.6	2363
San Andrés Tuxtla	27.8	20.1	2261
Tapalapan	27.1	21.7	2186
Coyame	26.4	20.0	4160
Catemaco	26.6	19.6	2045
Guasuntlan	27.7	19.7	1763
*Dirección General de Geografía y Meteorología; Comisión del Papaloapan (1958).			

Rainfall. Precipitation in the Sierra varies considerably in intensity, amount and monthly distribution. Since no data are available for the Gulf side of the mountains, it is not possible to do more than estimate the amount and pattern of rainfall for a large part of the range. Records from stations on the southern slopes show decreasing rainfall with diminishing elevation. This is probably true also on the Gulf-facing slopes from several hundred meters down to sea level. That these slopes and also the south side of the mountains at greater elevations receive a high annual rainfall, chiefly the orographic type, is certain from data for other tropical mountains in trade wind latitudes. Other than through personal experience of rainfall and observations of the increase in luxuriance and height of vegetation from sea level to higher elevations, the only indication of this greater precipitation is the data from Coyame. This station is located on the northeast shore of Lago Catemaco at an elevation of about 340 meters. It is the only meteorological station near the Gulf slope and lies slightly below the gap forming the lowest point in the ridge to the north. Thus it receives the orographic rainfall from the north and northeast winds which come in partly through the low lying pocket of the Bahía Sontecomapan. Its mean annual rainfall of 4160 millimeters is the highest recorded in the Sierra. Although this amount may be in part due to the local topography, it indicates that there is a high annual rainfall, probably more than 4000 millimeters, at a level above this station on the Gulf slopes of the mountains. Richards (1957:143) referred to work by Brown on Mount Maquiling in the Philippines in which the latter recorded a slight decrease in rainfall at 1000 meters elevation compared to that at 740 meters. Waibel (1933:50) gives data for orographic rainfall at stations above Tapachula in the Sierra Madre de Chiapas in which annual precipitation decreases above about 700 meters elevation. Probably a similar situation exists in the Sierra de Tuxtla. These mountains above 700 meters elevation are frequently cloud covered so that air at these levels on windward and leeward slopes is near or at the saturation point during part of the year. Table 1 shows that none of the other stations in the range approaches the rainfall amount received at Coyame, their locations on the inland side at fairly low elevations effectively removing them from the zone of highest precipitation.

There is a minimum of rainfall in March or April and maximums in July and October. The rainy season usually begins in June and continues into November (Figure 4). The decrease in precipitation which frequently, but not always, takes place in August (the short dry period called *el veranillo*) is due to the more stable air resulting from the intertropical front's retreat toward the south under influence of the growth and westward displacement of the Atlantic anticyclone (Garbell, 1947:118, 123). Wallén (1956:149, 152) showed that the Sierra has a low variability of annual precipitation (15-20 per cent), particularly its eastern section, compared to most parts of Mexico. Annual precipitation can vary considerably at a weather station. For example, in 1952 Catemaco (mean annual 2045 mm.) had 2806 millimeters, and in 1958, 2353 millimeters; and Coyame (mean annual 4160 mm.) had 5248 millimeters in 1952 and 5296 millimeters in 1958. Variability in monthly precipitation is highest from June to September, a period when the region is affected by the irregular occurrence of easterly waves, *el veranillo* and local convectional storms.

From February to May the driest period of the year occurs. Several stations, such as San Andrés Tuxtla with a yearly average of 2261 millimeters, in some years have no measurable precipitation in April. The severity of this dry season decreases with increase in elevation as well as with location toward the windward side of the mountains. The year 1962 was drier than usual and sugar cane did not do well. On May 28, during a normally wet period, I observed a large "dust devil" (5-10 meters in height) west of Cerro Mono Blanco. Such a phenomenon is uncommon in the region. The length of the dry season varies, the rains sometimes commencing a week or two early in May, or the latter part of the month of January being abnormally dry.

Rainfall is frequently heavy during severe northers from October to early April and in convectional storms from May into October. During the latter period light rains frequently begin by evening with the heaviest downpours after midnight or in the very early morning. At other seasons precipitation often occurs during the day. Rain storms are often accompanied by thunder and lightning, sometimes severe, from late April into November. Such displays occur mostly at night. At Playa Azul on October 27 and 28, 1962, there was almost uninterrupted rain of varying intensity for thirty hours (ca. 250 mm.) in a norther. On August 26, at the same locality, 78 millimeters fell in two and one-half hours from 11:30 AM to 2 PM. Occasionally light drizzle accompanies a low-lying stratus layer which extends to below 600 meters elevation. The diversity of terrain in the Sierra contributes toward varying amount and intensity of rainfall, especially where windward and leeward slopes are concerned. For example, I experienced heavy rain on the Gulf slopes of the ridge north of Lago Catemaco, although no rain fell from the same cloud layer over Playa Azul on the lake shore.

December, January and July are usually the cloudiest months, while April, May and August are the least cloudy. On many days throughout the year, however, the highest peaks are cloud covered even when it is clear over most of the Sierra. Fog is infrequent, occasionally occurring at upper elevations, but more often in the valleys on the southern slopes. There are occasional hazy days especially when winds are southerly and in the drier period when vegetation is being burned.

Northers. The cold air masses from the north which cross the Gulf of Mexico and contact the Sierra de Tuxtla occur from October to April. During my field work in 1960 and 1962 I recorded northers on an average of every 3.5 days in March and early April. This is a high frequency compared to the number in February, November and early December 1962. During January 1962, however, there were six northers, several being strong outbreaks of polar air. Waibel (1938:411-412) said that Tampico averaged 25.8 northers during the October-April period from 1914 to 1921, an average of 4.2 per month.

Northers in the Sierra vary considerably in intensity and duration, some being weak and of only one or two days duration, others lasting four or five days with pronounced effects on temperature and precipitation. Waibel (1938:411) stated that northers attain their greatest frequency, strength and extent in the months of December through February when thermal differences are greatest between high and low latitudes. This is true on the average especially in regard to low temperatures and southward reach, but not necessarily with respect to storminess, duration and rainfall. February 1962 had no northers as stormy or with as much precipitation as those I experienced in March 1960 and 1962.

The advent of a norther in the Sierra is characterized by a distinct temperature drop which may vary in relation to season and severity from as little as 4 or 5 to over 10 degrees Centigrade in 24 hours. Figure 5 shows the progression of mean daily temperatures recorded at Playa Azul and Veracruz city during five northers in early 1962. Playa Azul is in a sheltered location, as are some of the other Sierra stations. Therefore, temperature decreases would be greater on the Gulf side of the range. A comparison between temperature progression in the northers of February to April and the one in January shows a similar pattern in which temperature usually rises slowly or remains steady before a norther, reaching a maximum just before the cold front arrives. The minimum temperature occurs after the frontal passage, usually on the third or fourth day after the initial downward trend; it rises

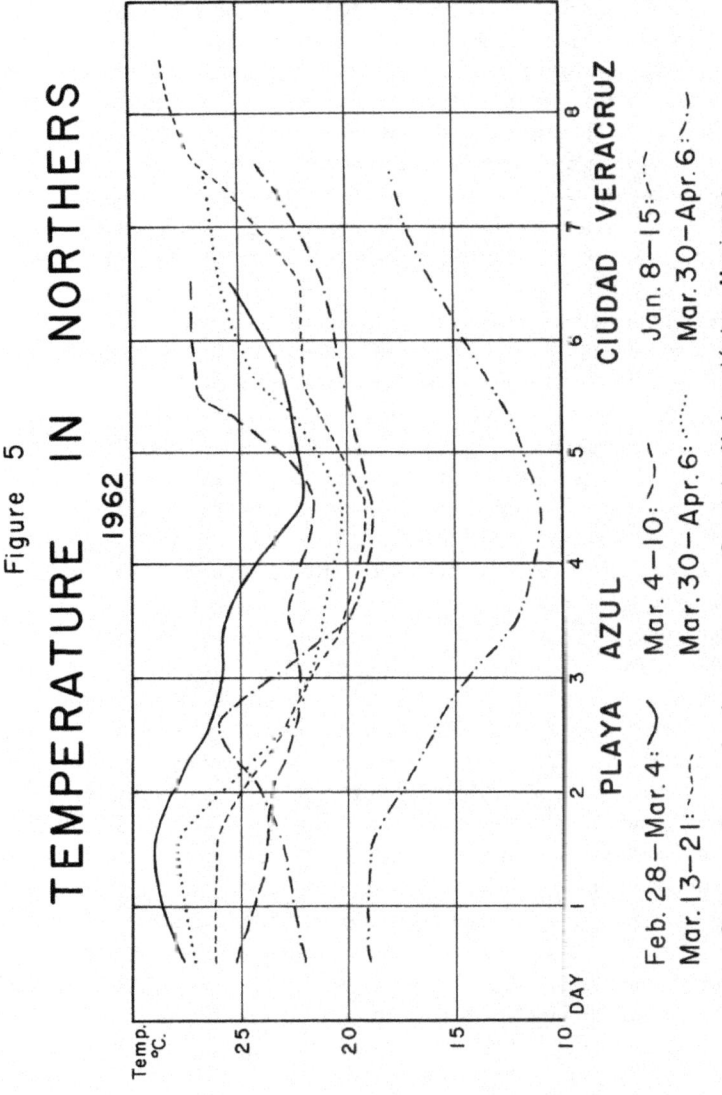

Figure 5

TEMPERATURE IN NORTHERS

1962

PLAYA AZUL

Feb. 28—Mar. 4:) Mar. 4—10: ⌐
Mar. 13—21: —·—

CIUDAD VERACRUZ

Jan. 8—15: ⌐
Mar. 30—Apr. 6: ····· Mar. 30—Apr. 6:)

Sources: Personal observations; Servicio Meteorológico Mexicano

more rapidly afterwards, reaching a peak on the second or third day. The minimum temperature I recorded during a norther was 9 degrees Centigrade on April 5, 1960, at 500 meters elevation above Dos Amates. Table I contains lowest minimum temperatures at Sierra stations, all recorded in December, January or February, probably in northers. Waibel (1938:417) mentioned that Veracruz city is occasionally affected by frost in northers; there are no such records at Sierra stations, but indications are that temperatures reach the freezing point at least at the higher elevations. Barometric pressure normally decreases slowly in a one or two day period before the norther's arrival in southern Veracruz. With the advent of the cold air, the pressure increases fairly rapidly at a decelerating rate. The increase is variable depending on the movement and depth of the anticyclone, and pressure changes range from less than 1 to over 10 millibars in 24 hours.

As the norther arrives the wind backs from south, southeast or northeast to north or sometimes north-northwest. Although initially the wind during a norther is from north or slightly west of north, it veers gradually later to north-northeast or northeast. Waibel (1938:416) referred to a similar wind change after one to two days in northers over Cuba. In the Sierra this wind shift is frequently accompanied by a marked increase in velocity, and though usually steady, the wind occasionally becomes gusty, this partially being the result of terrain configuration. On a ridge exposed to the north (500 meters a.s.l.) I recorded velocities of up to 80 kilometers per hour on March 18 and April 3, 1960, during strong northers; usually, however, the wind is only about 15 to 30 kilometers per hour and is considerably less after the first day. Waibel (1938:415) remarked on the increase in velocity of norther winds as the cold air moves south over the Gulf of México. The wind either remains light northeast or veers to southerly when clearing takes place after the norther.

In the period before northers the weather is usually fair and clear, or mostly clear with only scattered cumulus clouds. Altostratus and altocumulus clouds appear before the storm and as the front strikes the mountain slopes, heavy roll cumulus forms and initially towers over the windward side of the range, eventually enveloping the peaks. In a strong norther clouds move across the Sierra to the south forming a layer of stratocumulus and sometimes stratus with a ceiling down to as low as 400 meters. Occasionally clear sections occur where altocumulus clouds can be observed well above the lower layer. At the beginning of a norther the lower part of the ridge north of Coyame becomes the path for a low cloud layer moving southward over Lago Catemaco, although no clouds cross to the south side of the range elsewhere at these times. A gradual lifting of the cloud ceiling takes place as the norther wanes, but thick roll and towering cumulus remain against the Gulf slopes after partial clearing occurs inland.

Precipitation in northers is often heavy downpours of brief duration during the day on the Gulf slopes and peaks of the Sierra and on south slopes at high elevations. Precipitation is usually lighter on the lee side, and in some northers I noted only a fine mist or drizzle south of the crest of the range. Night rainfall occurs frequently during northers, and is more prolonged and lighter than in the day. At times I experienced no rain in the lowlands to the south and west of the Sierra when considerable rainfall was occurring in the mountains. I seldom recorded more than 5 or six millimeters of rain at Playa Azul in a norther, although undoubtedly more fell on the Gulf slopes.

The daily weather maps for 1962 show that the Sierra is usually contacted only by the southwest quadrants of the anticyclones which bring northers south and southeastward across the United States. Their centers seldom move south of Texas into Tamaulipas, and subsequently travel eastward across the southern United States or northern Gulf of Mexico. Consequently, the western sections of their cold fronts contact the Sierra de Tuxtla with a north or often northeast wind component. The isobaric pattern, especially in the stronger northers, shows a bellying southward in the Gulf with the apex considerably east of the Sierra toward the Yucatan Peninsula. Thus the most severe cold and wind conditions normally do not affect the range. Brands (1944:115) and Waibel (1938:415) mentioned that the northers often arrived at Frontera, Tabasco before reaching Tampico, and both of these localities before reaching the city of Veracruz. This points out the Sierra's position in a pocket west of the southern apex of the Gulf where it frequently escapes the storm's full force.

The cold air mass of the norther is warmed in passing over the Gulf waters before striking the Sierra (18°30' N). Its moisture content is increased and the orographic uplift of this modified air as it strikes the Sierra causes a rainfall which is higher than at most lowland coastal areas. Waibel (1933:424) said that 55 per cent of the over 4000 millime-

35

ters of rain annually at Teapa, Tabasco, falls from October to March. This station is at the base of the Chiapas mountains and, although there are no stations on the Gulf slope of the Sierra, it is interesting that about 50 per cent of the rainfall at Coyame, the station nearest the Gulf side, occurs in the same period. Much of this precipitation falls in the northers so they are a significant factor in affecting the moisture conditions prevailing in the Sierra. The creation of suitable conditions for the growth of humid evergreen forests on the Isthmus of Tehuantepec was attributed by Waibel (1938:419) partly to the heavy precipitation in the northers. He also mentions that these thick forests help to break the force of the northers, a function which the extensive forests on the Sierra's slopes also perform.

Easterly waves. The Sierra de Tuxtla's climate is influenced by waves in the easterly winds probably most often from June into October. In 1962 I noted several weather disturbances which were partially the result of such low pressure troughs moving westward from the Caribbean Sea. Their frequency and the degree to which they affect the region are variable, partly because the Sierra is located west of the location (80-85°W) where they often disintegrate (Riehl, 1954:222). With the limited time and equipment available during my field work, it was not possible to obtain detailed observations of weather conditions to distinguish a wave disturbance from the usual rainy season convectional storms beginning in May. Riehl remarked (1954:222) that areas of bad weather in easterly waves would lie west of the trough line when the upper air is warmer in the subtropics than the trades, and that the bad weather zone tends to be displaced westward relative to the troughs as the wave speed increases. This would bring the effects of waves closer to the Sierra even if disintegration took place, especially since the waves are often 1600 to 3200 kilometers in breadth and last several days (Riehl, 1954:211).

From June 10 through 16, 1962, a moderately intense easterly wave contacted the Sierra and caused bad weather. It is possible to locate this wave from the isobar pattern on the weather maps as it moved in from the Caribbean Sea. From June 12 through 17, rainfall occurred in Yucatan and the Isthmus of Tehuantepec in connection with this disturbance. Possibly the Sierra did not receive the full impact of the bad weather associated with this wave because the latter appears to have weakened somewhat to the east of the range. At Playa Azul a slight lowering of the daily average temperature occurred (1.7 degrees Centigrade from June 9 to 13) during the wave passage. This was caused by the stronger northeast wind, increased day cloudiness, rain and radiation on the clear nights of June 13 and 14. Pressure at Playa Azul dropped about 5.1 millibars in the same period. Records at Coyame showed traces of precipitation on June 9 and 10 and 13 through 16. On June 11 this station reported 14 millimeters and 4 millimeters the next day. San Andrés Tuxtla reported a trace of rain on June 11, 1.5 millimeters on the 12th and 7 millimeters on June 13. These rainfall records indicate the gradual wave progression westward and correlate with my own observations of its movement and precipitation associated with it.

An account follows of the weather conditions I recorded before, during and after the wave passage.

June 8 - fair, clear, later in day Cu clouds with moderate vertical development; wind S 10-25 kph, later S 15-40 kph; a fair, calm night, mostly clear with lightning far to ESE.

June 9 - fair, partly cloudy with Cu clouds developing and some Ci, by midafternoon towering Cu clouds to E, S and SW with heavy Cu roll clouds against mountains from the Gulf; wind now NE 15-30 kph; a breezy, partly cloudy night with lightning to S.

June 10 - fair, small Cu clouds dissipating; wind NE 5 kph, later NE 25-50 kph with Cu clouds and some vertical development, trace of precipitation; night cool, clear with lightning to NE early.

June 11 - fair early with heavy clouds and rain to NE over Gulf, later Cu clouds developing over Sierra; wind NE 25-50 kph; at first a breezy night, later calm, cloudy; about 2 AM a brief, heavy rain (3 mm.).

June 12 - fair early, mostly cloudy, dark to north, becoming overcast with light rain beginning 6:30 AM. Sc, Cu with vertical development, and some Ac; wind NE 10-40 kph, gusty; a calm, cool night.

June 13 - fair early, overcast, Sc clouds with rain beginning 5:45 AM, clouds low over Gulf slope where light, then heavy rain showers; later in day thunder, As, Ac and Ci clouds also, with heavy Cu to south of Sierra, a chaotic sky; wind NE 15-40 kph, variable; a calm, cool, clear night.

June 14 - fair, calm at first, some Cu over Gulf slope, a chaotic sky with Cu, As, Ci and Cu Ni to S; wind NE 10 kph; a fair, clear and calm night.

June 15 - fair, partly cloudy, heavy Cu roll against Gulf slope; wind NE 10-15 kph; a fair, mostly clear night.

June 16 - fair, in early morning Cu and As clouds; wind SW 10 kph; a heavy rainstorm with thunder and lightning moved in from the south at 6:30 PM marking the end of the wave passage and a beginning of a cloudy period of low pressure and southerly winds in conjunction with the deepening and movement southeast of the low cell over north-central Mexico.

The change of wind in an easterly wave from north to south was noticeable in the passage of this and other waves over the Sierra. During the wave's passage the Atlantic high cell shifted south, but appeared to weaken in the western Caribbean Sea, the north central Mexico low cell intensified and lower pressure moved northward across the Isthmus of Tehuantepec. It appears that the barrier presented by the Sierra is an important factor in increasing the turbulence and rise of the unstable air in easterly waves with consequent increase in storminess and precipitation.

Hurricanes. Only four hurricanes have come close to or struck the Sierra de Tuxtla since their courses have been charted. The dates of occurrence and the tracks of these four are shown in Figure 6. This also shows the courses of other storms of hurricane intensity which have passed through the southern Gulf of Mexico. The Sierra's location is far removed from the path of greatest frequency of tropical cyclones, which is through the northern Caribbean Sea. Most storms in Figure 6 have kept on a westward or west-northwestward path and have not recurved, as most do, to the northeast. This is probably due to the weakening of the upper air trough to whose northeast wind flow they normally respond, to the disappearance of the temperature gradient necessary to maintain the upper westerlies in the southern portion of the trough, and a consequent replacement of these west winds by easterlies which carry the cyclone westward (Riehl, 1954:348-349). Occasionally hurricanes crossing Yucatan intensify as they move into the Gulf of Campeche where conditions are often conducive to genesis of new storms. The hurricane of August 31 to September 8, 1888, and the others with a similar direction. probably moved southwest in response to a deep northeasterly flow. According to Riehl (1954:351), this is a type of upper air current common in the western Gulf, especially in August.

There is little information about the influence of hurricanes on the Sierra de Tuxtla. Tannehill (1939:222) listed two tropical storms for "Vera Cruz" on November 1 and 26, 1838, but included no more data. He also recorded (1938:241-242) two in the Bay of Campeche on September 25-27, 1892, and on June 12-30, 1893, with no further information. Concerning the 1888 hurricane mentioned above, Tannehill (1938:160) said that it was a severe one and struck the Mexican coast south of Veracruz city on the night of September 7-8 doing "great damage." On one map this hurricane is shown striking the northwestern part of the Sierra, but Fassig (1913:Plate I) depicted it hitting the coast north of Alvarado and closer to Veracruz city. Garriott (1900:36) said it struck the Mexican coast between Veracruz and Coatzacoalcos at about 18°S 97°W, a point actually not on the coast but about 60 kilometers inland. Medel y Alvarado (1963:[1]321) stated that this hurricane killed 50 per cent of the fauna and left the rest in need of trees for shelter and fruits for food. It destroyed a "great number of huts, principally those of the higher elevations." He also stated that numerous human lives were lost by drowning in the rural areas. Hit especially hard was the southern part of the region suffering the loss of "hundreds of human lives," cattle and crops. Medel y Alvarado (1963:[1]441) spoke of a repetition of this destruction in a cyclonic disturbance on September 23, 1898, which brought heavy rain causing numerous deaths by drowning. The level of Lago Catemaco rose two meters; and flooded arroyos caused the isolation of San Andrés Tuxtla. "Fifteen years afterward," he continues, "still there could be seen through all parts, visible vestiges of the great

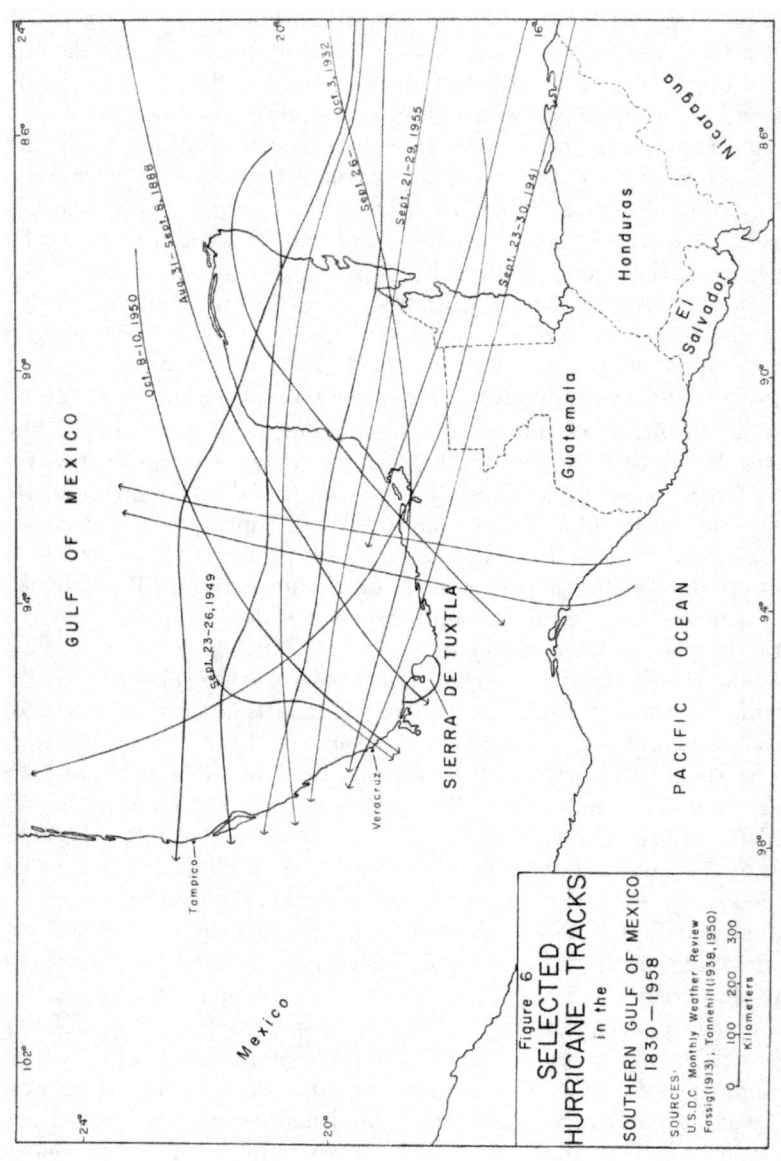

Figure 6

SELECTED
HURRICANE TRACKS
in the
SOUTHERN GULF OF MEXICO
1830-1958

SOURCES:
U.S.D.C. Monthly Weather Review
Fassig(1913); Tannehill(1938,1950)

0   100   200   300
Kilometers

GULF OF MEXICO

Mexico

Tampico

Veracruz

SIERRA DE TUXTLA

Oct. 8-10, 1950

Aug. 31-Sept. 8, 1888

Sept. 23-26, 1949

Sept. 26-

Oct. 3, 1932

Sept. 21-29, 1955

Sept. 23-30, 1931

Guatemala

El Salvador

Honduras

Nicaragua

PACIFIC OCEAN

and numerous landslides on the cerros." This storm's track was not shown on maps.

The storm of September 23-26, 1949, hit the coast at Veracruz city with maximum winds recorded at 128 kph. The hurricane of September 26 to October 3, 1932, was a severe one, causing great damage in Puerto Rico, and Medel y Alvarado (1963: 2]281) mentioned strong winds and inundations on October 5, 1932, causing considerable damage to banana plantations in El Laurel at the southwestern base of the range. Hurricane "Item" of October 8-10, 1950, hit the coast a short distance south of Veracruz city with 175 kph winds. Norton (1951) reported that it caused "heavy" damage but there was no estimate of casualties. The hurricane of September 23-30, 1941, heavily damaged crops in Central America (Sumner, 1941a:363) but decreased rapidly in intensity as it moved into the Gulf and went inland near Veracruz city as a week depression (Sumner, 1941b:266). It is of interest that "Dora," a storm of doubtful hurricane intensity struck the Veracruz coast on September 12, 1956, near Tuxpan, with winds of only about 60 kph, but dropped heavy rain in the vicinity causing deaths from landslides and flooding. Dunn, Davis and Moore (1956:440) mentioned a news dispatch reporting seven persons drowned in the overflowing Pachuapan River near San Andrés Tuxtla in connection with this storm. Most of the hurricanes in the southern Gulf cross Yucatan and strike the Mexican coast far northwest of the Sierra. Sumner (1944:239), Norton (1952:3) and Dunn, et. al. (1955:322) reported considerable flooding, deaths and crop destruction in the Isthmus of Tehuantepec, Yucatan and the Tampico area from three of these hurricanes. One of them, hurricane "Janet" of September 21-29, 1955, is said by Argudin (1962:11) to have caused some forest destruction in the Sierra de Tuxtla.

From the known variability in hurricane size (100 to 1000 km. radius) (Riehl, 1954:282,284) and the fact that the centers of most of those shown in Figure 6 passed within 300 kilometers of the Sierra de Tuxtla, it is probable that the region was affected to varying degrees by a majority of them. Several residents of the Sierra told me of hurricane winds and damage to their houses at higher elevations, but I did not observe any recent destruction from these storms. The region has been on the left side of almost all the hurricanes, and in these two quadrants wind velocities, tide height and precipitation amount are usually markedly less than in the right ones (Cline, 1926:210,239).

Storm tides may at times have caused damage and loss of life in the low lying settlements on the coastal side of the Sierra. The inhabitants of these places, however, can quickly and easily reach high terrain to escape inundations. Thus, flooding from hurricanes has probably caused the most loss of life in the Sierra. Such occurrences have been usually restricted to low areas and along rivers and arroyos because of the generally high terrain gradient, the many streams to carry off flood waters and the good absorptive ability of the extensive forested sections.

1.4 Soils

No detailed soil studies have been published for the Sierra de Tuxtla. The articles by Ortiz Monasterio (1955-1957) and Macías Villada (1960) on Mexico's soils are general in nature. Soils in the Sierra are principally derived from volcanic materials. Since much of the region is forested, organic material from this source has contributed to soil materials over large areas. Surface soils range in texture from loamy sand to clay, the majority appearing to be clays (more than 40 per cent clay) or clay loams in which the clay fraction is the nonsticky type often found in warm climates (M. G. Cline, personal communication, Dept. of Agronomy, Cornell University). Ortiz Monasterio (1956) shows the soils of the Sierra to be medium to heavy textured clays and loams. The subsurface materials are variable and often deeply weathered. In many areas, particularly at the higher elevations, they are mainly layers of ash, scoriae, fractured lava and scattered larger fragments of lava, tuff, basalt, mineral nodules and other igneous rocks (Plate X). Soils derived from volcanic materials, especially those rejuvenated by more recent ash falls and formed from highly basic basalt, are the most fertile. They contain a good supply of plant nutrients in the form of soluble mineral components including feldspars (ash particles) and oxides of iron, magnesium, potassium and aluminum.

In some places on the south slopes the soils appear to be derived from argillaceous rocks, mainly sandstones, fragments of which occur at and below the surface. Plowed soils often exhibit a fine powdery texture when dry. In many sections on both massifs where lava flows and layers of volcanic materials occur, soil horizons are indistinct or absent. More mature soils with better profile development occur in the older, much weathered material of the broad ridges on the southern slopes of the range. The varying depths of soils are apparent in road cuts, on ridge crests, and along deep ravine slopes and floors. In places the soil layer may be over two meters deep, but on some ravine floors, ridge crests and crater walls it may be only a few centimeters thick or absent. Outcrops of tuff and basalt are widespread.

In ten surface soil samples selectively collected in various localities in the Sierra, the organic matter content ranges from 1 to 10.7 per cent. The drop in organic matter amount from the forested area samples to those from cultivated or grazed sections is marked. The reaction of these topsoil samples varies from strongly to slightly acid (pH 5.0 to 6.2) (M.G. Cline, personal communication). Ortiz Monasterio (1956) on the soil map of Mexico, shows the Volcán San Martin Tuxtla massif as having neutral (pH 6.5-7.5) soils and the Santa Marta massif slightly acid (pH 6.4-6) soils. As the precipitation in the Sierra is heavy (1500 to over 4000 millimeters annually), it is to be expected that the soils are being leached to a considerable extent. The samples from the humid tropical forest as well as those from pine forest and oak forest show a low supply of extractable calcium and potassium (nutrients available for plants) based on soil standards for New York State. It was the opinion of an engineer with a fertilizer company that the soils of the Sierra are decidedly deficient in phosphorus available to plants, an expected condition where strong weathering is taking place. Foster (1940) mentioned "tierra roja" soil as the dominant type from Cuilonia to Ocotal Grande on the southern slopes of the Volcán Santa Marta massif. Some of the other ridges south and southwest of Cerro Cintepec also possess this reddish soil, but through most of the range soils vary from a dark reddish or grayish brown to dark brown or black.

Soil erosion in the Tuxtla mountains is principally sheet type. There is comparatively little gullying or slumping in evidence. The rapid growth of ground cover plants both in destroyed forest areas and in open fields effectively impedes excessive sheet erosion in most places despite the steep slopes. During heavy rain there is a high water runoff in some open areas and at times large amounts of soil and debris are washed into the streets of the towns from the surrounding trails and bare areas. Trails into the mountains from the larger towns are sometimes cut two or more meters below the land surface by long

43

Plate X. Subsurface Strata

.3 km. west Tapalapan, 500 m. a.s.l. Many alternating layers of volcanic ash, lava, gravels and conglomerates form thick deposits in parts of the Sierra. They are seldom visible unless exposed in road cuts such as this. The heavily forested volcano Cerro Blanco is at upper left.

usage and erosion. Soil porosity is high in many parts of the range, particularly in the extensive humid forest and at upper elevations where deep layers of ash and volcanic gravels exist. I found that the soil in a small tract on which the primary forest had recently been destroyed (one to two years ago) was very absorbent in a heavy downpour. There the soil is composed of ash and gravels and the slope not over 20 per cent. Sheet erosion seems greatest in the more open pine and oak forest areas on the south side of the Volcán Santa Marta massif where there is less vegetation to interrupt rain and retard runoff. The Soil Erosion Survey of Latin America (United Nations Conservation Foundation and the Food and Agricultural Organization, 1954), on a small scale map of Mexico, showed moderate erosion on the southern slopes of the Sierra and slight or none over most of the forested parts.

Chapter 2

FLORAL ELEMENTS

2.1 Introduction

The Sierra de Tuxtla is near the northern limit of the tropical region which extends from South America through Central America into southern Mexico. Its plant life is basically tropical in nature; the range is in the coastal lowlands which continue into Central America, and this connection, in conjunction with essentially warm temperatures and high precipitation, is conducive to the development of luxuriant vegetation with most components having Central and South American affinities. As in other mountainous parts of southern Mexico, however, temperate North American elements exist, locally constituting a conspicuous part of the flora (cf. Miranda and Sharp, 1950). The length axis of the Sierra in relation to prevailing winds results in a rain shadow on the southern side, where some plants characteristic of subhumid tropical regions have developed. Volcanism and differential weathering have caused a diverse topography in the Sierra. This surface configuration plus changes in soil depth and moisture, and the variations among soils derived from recent and older volcanic materials and from sedimentary rocks result in not only different plant formations but marked structural complexities within formations. Such variability is increased in many parts of the Sierra by human disturbance resulting in all stages of secondary succession.

The extent of the region and the nature of my investigations permitted only a general survey of the plant cover. From ground observation and aerial photographs I have been able to determine approximately the areal distribution of some principal plant formations. This is shown in Figure 7. Not depicted is the transitional character of the boundaries between formations. I also selectively collected plant specimens in many parts of the Sierra. I restricted collecting to almost entirely woody plants, to those which were obviously dominants in each formation or, where a multiplicity of rain forest species existed, to those which appeared to be numerous as evidenced by fruits or flowers. Appendix B contains a list of the trees, shrubs and herbaceous plants identified.

2.2   Plant Formations

It is possible to distinguish seven forest formations in the Sierra. The extent of open fields with tree rows and thickets is considerable. The Savanna, Mangrove and Littoral formations occupy small areas at low elevations. A list of the ten types follows.

1. Rain forest

2. Cloud forest

3. Lowland valley and swamp forest

4. Semideciduous forest and palm-grassland

5. Pine-oak forest

6. Oak forest

7. Gum-oak forest

8. Savanna

9. Forest remnants, tree rows, thickets and open fields

10. Littoral and mangrove

Rain forest. Most of the Gulf-facing slopes of the Sierra are covered with an essentially evergreen humid tropical forest occupying the zone of highest rainfall. This type of forest also occurs on the southern

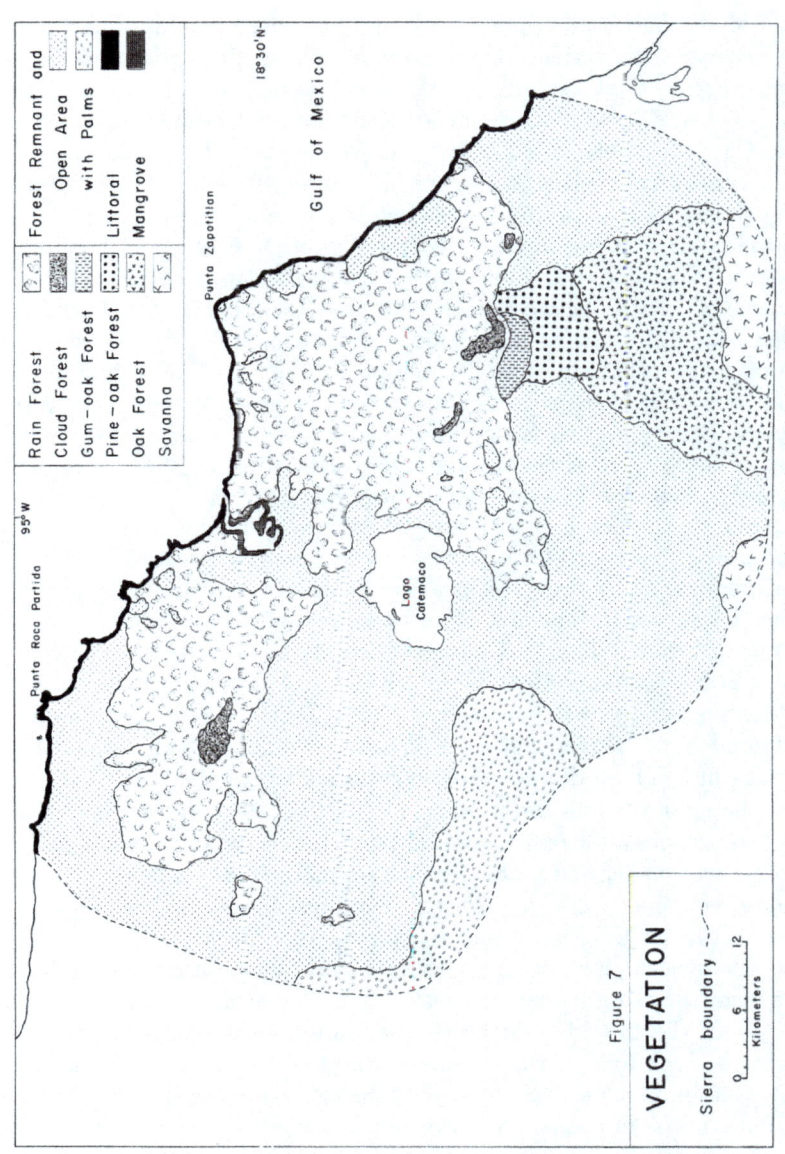

Figure 7

VEGETATION

Sierra boundary ----

0   6   12
Kilometers

Legend:

Rain Forest
Cloud Forest
Gum-oak Forest
Pine-oak Forest
Oak Forest
Savanna

Forest Remnant and
Open Area
with Palms
Littoral
Mangrove

Punta Roca Partida

95° W

Punta Zapotitlan

18° 30'N

Gulf of Mexico

Lago
Catemaco

side of the Sierra at upper elevations, but there it is limited due to decreasing precipitation, edaphic conditions and clearing for agriculture and grazing. The most extensive areas remaining south of the high length axis of the range are in the vicinity of the Cumbres de Bastonal and Cerro Cintepec and southwest of Volcán San Martín Tuxtla.

Primary rain forest in the range is variable in structure. This is due to amount of rainfall, soil conditions and slope exposure, and especially to the variable land configuration. It is likely, although no data are available in the Sierra, that the highest precipitation falls at middle elevations between about 600 and 1200 meters (cf. Richards, 1957:142-143). The primary rain forest at these altitudes seems to be tallest, most luxuriant, with the greatest proportion of large trees and a more definite stratification. Yet, in other places, such as on Cerros Buena Vista, Blanco and Tuxtla, the primary forest at 700 meters is low, sometimes stunted and contains few large trees; in such localities, however, it is on steep slopes near the peaks and is frequently exposed to the north, a situation which may partially account for its character. The sword-leafed cycad *Ceratozamia mexicana* Brogn. was conspicuous on the upper slopes and peaks of Cerro Blanco, the only place I observed it. The only other cycad recorded was *Zamia loddigesii* var. *angustifolia* (Regal) Schuster, found by Gary Ross in the ground cover of forests above 1000 meters on Volcán Santa Marta. A common understory plant at the summit of Cerro Tuxtla was the red-flowered *Odontonema callistachyum* S. & C. Kuntze, which averaged about two meters in height; I did not observe it so numerous elsewhere.

The montane rain forest is usually best developed on ravine slopes and on broader inter-ravine areas (Plate XI); the crests of narrow ridges often have smaller and lower trees with a more dense understory. Some wide ravine floors also possess well developed forest. Soil is frequently thinner on ridge crests than on slopes, and rocks are occasionally exposed in such places resulting in less favorable conditions for moisture retention and high forest growth. The decreasing luxuriance and height of the humid rain forest is apparent in traveling southward from the high axis of the range. Epiphytes, climbers and lianas are not so abundant and large trees less numerous, reflecting the change in moisture conditions. At lower elevations on the Gulf slopes the montane rain forest still exhibits the variable character mentioned above, but also shows changes similar to, though less pronounced than,

50

Plate XI. Montane Rain Forest

5 km. south-southeast Colonia Huatusco, 650 m. a.s.l. This humid evergreen forest covers a large part of the Sierra. Lianas, climbers, epiphytes and tree buttressing are characteristic. The spiny palm *Astrocaryum mexicanum* Liebm., visible at lower center, is a typical constituent of the understory.

those evident on the southern slopes.

In the tall, well developed montane rain forest there are three tree strata. On the forest floor, which is covered with a shallow leaf litter, and on ravine banks, low herbaceous plants such as the red-flowered *Aphelandra aurantiaca* (Scheidw.) Lindl. are numerous; there are few ferns here, *Didymochlaena truncatula* (Swartz) J. Smith being most numerous, with also specimens of the tree ferns *Cyathea* sp. and *Alsophila schiedeana* Presl. There are small saplings and shrubs two to five meters tall, such as *Hamelia longipes* Standl., *Myriocarpa longipes* Liebm., *Cephaelis elata* Sw. and species of *Piper*, scattered through the forest, but most prominent is the spiny palm *Astrocaryum mexicanum* Liebm. and in lesser numbers the palm *Chamaedorea tepejilote* Liebm. Most of these palms and shrubs are less than five meters tall.

In many places it is not difficult to walk upright in this open understory. The middle stratum is composed of small and medium sized trees about 6 to 18 meters in height. Some of these are *Pseudolmedia oxyphyllaria* Donn. Sm., *Stemmadenia galeottiana* (A. Rich.) Miers, *Pleuranthodendron mexicana* (A. Gray) L. Wms., *Calatola* sp. (possibly *C. laevigata* Standl.), *Chlethra macrophylla* M. & G., *Saurauia* sp., *Annona* sp., and *Coccoloba* sp. (possibly *C. montana* Standl.). This last tree is called *uvero* by the local people. I also found much larger specimens of *Pseudolmedia* and *Clethra* within the forest canopy. Although some of the largest trees might be considered part of an emergent stratum, in most cases they are not numerous enough nor do they protrude sufficiently to be considered separate from the third or highest stratum of trees which forms the canopy. The latter consists of trees varying from 20 to 35 meters in height, with few exceeding 35 and most about 25 to 30 meters. Diameter at breast height ranges from about .75 to 1.75 meters with some trees exceeding 2 meters. Density of the larger canopy trees changes with the terrain variation. Very large trees may be located at rather widely spaced intervals in the lower montane forest and in places at upper elevations.

The highest stratum is composed of a variety of tree species. Several were common in many parts of the Sierra, their flowers or fruits being conspicuous at various seasons. Some canopy trees are: *Bernoullia flammea* Oliver, *Talauma mexicana* (DC.) G. Don (called *llolo* by Mexican assistant), *Pithecollobium arboreum* (L.) Urb., *Mirandaceltis monoica* (Hemsl.) Sharp, *Phoebe mexicana* Meiss., *Sloanea* sp. and various species of *Ficus*. Also in the upper stratum I found *Engelhardtia guatemalensis* Standl. and *Virola guatemalensis* (Hemsl.) Warb. These may be the first recorded for Mexico. *Dussia mexicana* (Standl.) Harms, called *jaboncillo, Ilex discolor* Hemsl. and *Ilex* sp., possibly *I. tolucana* Hemsl. (called *palo verde*), are also canopy trees in this rain forest. Some large oaks are present in the montane rain forest near the Cumbres de Bastonal and on Volcán San Martín Tuxtla. These are *Quercus skinneri* Benth. Davis (1952:314) described a coffee finca south of Lago Catemaco in which trees vary from one to eight feet (.3 to 2.4 meters) in diameter and were 60 to 70 feet (18.4 to 21.4 meters) high. Of the numerous tree species present, he mentioned *Elaphrim* [*Elaphrim* = *Bursera*] *simaruba, Mammea americana, Guazuma ulmifola, Ficus glaucacens* [*glaucescens*], *Tabebuia pentaphyla, Lonchocarpus guatemalensis, Ceiba pentandra* and *Cecropia mexicana*. He reported vines and epiphytes as common. Trees of the genus *Ficus* are fairly numerous in the region; they occur as canopy trees in the rain forest, along lake and stream edges and in tree rows bordering fields. Provisional determinations of some large *Ficus* in the rain forest are *Ficus kellermannii* Standl., *Ficus segoviae* Miq. (or *F. podifolia* H. B. K.) and *Ficus tecolutensis* (Liebm.) Miq. (possibly *F. cookii* Standl.). *Ficus glaucescens* (Liebm.) Miq. was found along the shore of Lago Catemaco, *Ficus cotinifolia* H.B.K. in tree rows and possibly *Ficus radula* Willd. along streams at lower elevations.

Strata in the montane forest at lower elevations and in some topographic situations at higher altitudes are less distinct, sometimes reduced to two, or are not apparent. Vines, air plants and climbers are present in abundance on the limbs and trunks of most medium and large sized trees in the primary rain forest. Buttressing is fairly common on such trees. Stilt roots occur sporadically, but are usually confined to medium and small sized trees more often in secondary forest. There is a continual fall of leaves, flowers and fruits through the year, especially during periods of heavy rain and strong wind. There did not seem to be any pronounced fruiting or flowering season, although I observed more leaf fall and flowering during the drier period and toward its end than in the rainy season. I did not observe any trees of the montane rain forest totally lacking leaves during the dry season except for *Bernoullia flammea*. Some of the trees in the open and in forest edge, however, lose most of their leaves. The canopy of some forest sections, when viewed from above, showed a distinctly greener aspect with the advent of the rains in May and June.

The taller rain forests of the Sierra have characteristics of a tropical forest formation mentioned by Beard (1944a:138). They possess the palm understory, general structure, height and often the three tree strata of his "Evergreen Seasonal Forest," although he classes this as a lowland (below 760 meters) formation; the montane forest at lower altitudes in the Sierra frequently does not show the two strata and lack of well developed buttressing characteristic of Beard's "Lower Montane Rain Forest," nor are palms an unimportant component as in his type. Although many of the large trees in the taller rain forest of the Sierra possess long, clean boles of 20 or more meters, there are also many which do not, particularly at lower altitudes. I would class the rain forest of this region as generally a subformation modified by terrain variations, soil conditions, and by climatic elements, with the dry season probably one of the important influences.

Cloud forest. The transition from tall rain forest to a lower type with predominantly medium sized trees occurs gradually and irregularly above about 1000 to 1200 meters on the four largest volcanoes. Since the mountains do not rise above the Upper Tropical Zone, the cloud forests of this zone cap the high peaks and occur on their southern slopes. Trees of the closed canopy are usually less than one meter in diameter and the forest averages about 20 meters or slightly less in height. There are generally two tree strata, a dense understory, many epiphytes and considerable amounts of moss on limbs and trunks (Plate XII). The palms *Astrocaryum mexicanum* Liebm. and *Chamaedorea tepejilote* Liebm. are present, but are less numerous than at lower elevations. *Clethra suaveolens* Turcz., *Oreopanax xalapense* (H.B.K.) D. & P., possibly *O. capitatum* (Jacz.) D. & P. and *Xylosma* sp. are some of the trees I found in this forest type. Gum, of North American affinity, *Liquidambar styraciflua* L. also occurs, but it does not appear to be numerous. Although the tree fern *Cyathea* sp. is common, it is not well distributed and is often restricted to ravine slopes and bottoms in the more moist places. I found large examples of this fern in such moisture pockets down to 550 meters on the Gulf slopes of the range, so it cannot be considered a characteristic species of the high cloud forest in the range. Like the taller montane forest at lower

Plate XII. Cloud Forest

Volcán San Martín Tuxtla, 1325 m. a.s.l. This type of forest occurs high on the large volcanoes. It is characterized by medium sized trees and a dense understory of palms, shrubs and herbaceous plants. Large tree ferns, such as the one at right center, also occur, and the limbs and trunks of trees are partly moss-covered.

altitudes, cloud forest is a variable formation, its chief characteristics being lower height, smaller trees and more dense undergrowth of woody and herbaceous plants. Species of *Piper* are common in the understory.

The crater walls and peak ridges of the big volcanoes possess an elfin or mossy woodland composed of small, many branched, gnarled trees, mostly 3 to 10 meters tall, their limbs and trunks heavily covered with mosses such as *Pterobryum densum* (Schwaegr.) Hornsch. and *Pilotrichella flexilis* (Hedw.) Jaeg.; they also bear numerous orchids and small epiphytes (Plate XIII). These trees are usually closely spaced, but most possess small leaves so that light penetrates well. There are few palms. The ground cover consists of lichens, mosses, shrubs and many low herbaceous plants with occasionally grasses in the lighter and more moist situations. Soil is usually thin and rocks are often exposed. Some of the tree and shrub species are *Clusia salvinii* Donn. Sm., *Senecio* sp. (possibly *S. schaffneri* Sch. Bip.), *Hoffmania lenticillata* Hemsl., *Viburnum acutifolium* Benth. and *Ilex nitida* (Vahl.) Maxim. The last is a conspicuous small tree of the elfin forest. *Oreopanax xalapense* (H.B.K.) D. & P., attaining larger size at lower elevations, is also one of the small trees present in this forest type. This formation is more restricted in area on the volcanoes of the southeastern massif than it is on Volcán San Martín. Gary Ross reported that the tree *Podocarpus oleifolius* D. Don is numerous in the elfin forest on the crest ridges of Volcán Santa Marta. The clouds frequently enveloping the forest are usually the result of condensation in rising moist air moving in on north and northeast winds from the Gulf. They often cover the volcanoes to well below the 1000 meter level on all sides, thus contributing to the moisture supply of the tall montane forest as well.

Lowland and swamp forest. This formation occupies nearly level terrain at low altitudes near the Gulf of Mexico along the valleys of several large streams between the volcanic ridges and about Bahía Sontecomapan. Because of its small extent it is not shown on Figure 7. It is composed of an upper story of medium to small sized trees mostly under .75 meter in diameter and from 15 to 20 meters tall. The understory is fairly open and composed mainly of small trees and palms, chiefly *Chamaedorea tepejilote*. Herbaceous plants are not numerous on the forest floor and are absent in many locations. Occasionally vines and saplings make an almost impenetrable cover. Pronounced buttressing is uncommon except in some larger trees along the water,

Plate XIII. Elfin Forest

Volcán San Martín Tuxtla, 1600 m. a.s.l. Stunted forest is characteristic of the highest ridges and crater walls of the large volcanoes. Mosses and orchids are abundant on trees, and herbaceous plants are common in the ground cover. Birds and mammals are not numerous in this habitat.

such as *Pachira aquatica* Aubl. During the rainy season parts of the lowland forest along streams may be inundated or have shallow pools, and sections bordering the mangrove fringe in the Bahía Sontecomapan may have a more or less permanent water layer.

Semideciduous forest. Much of this formation has been destroyed in the Sierra so that only remnants and more extensive sections on ravine slopes remain. It is distributed over the southern slopes of the range in the rain shadow areas where precipitation is usually less than 1800 millimeters annually. The formation is not shown in Figure 7 because of its disruption and small extent. Some of the largest tracts still intact are on the ridges and slopes south of Cerro Cintepec. Plate XIV shows this forest in June when new leaves are appearing on the trees. *Bursera simaruba* (L.) Sarg. is a common tree species. It is called *jeote* or *mulato* by the people in the Sierra. The mature forest of this type rarely exceeds about 25 meters in height. Tree trunks and limbs are angular with main branches often occurring low on the boles. The understory is dense with saplings, shrubs and occasional palms.

Buttressing is infrequent and few epiphytes, climbers or lianas are present. In the southwestern part of the range at lower elevations somewhat open mixed vegetation exists (Plate XV) which is considerably modified by man. Here the most prominent feature is the palms, mainly *Sabal* and *Orbignya* species, that grow singly or sometimes in dense groves. Coarse grass grows in the open areas often bordered by dense thickets of varying height. Patches of semideciduous forest in various stages of succession, tree rows and single trees are scattered throughout. Burning and grazing are important influences contributing toward expansion of grassland and reduction of woody plant growth.

Pine-oak forest. On the southern approaches to the Volcán Santa Marta massif this formation occupies the ridges and upper slopes of the ravines over about 100 square kilometers. Elevation limits extend from about 500 to over 1200 meters on the southeast ridge of the volcano. This is an open woodland with the trees spaced five or more meters apart (Plate XVI). *Pinus oocarpa* Schiede is the dominant species and possibly the only one here; individuals of about one-half meter in diameter are numerous and trees range from 15 to 25 meters in height. The ground cover is a coarse short grass, sparse or absent

58

Plate XIV. Semideciduous Forest

1.5 km. southwest Barrosa, 150 m. a.s.l. This drier forest occurs on ridges and ravine slopes on the southern side of the Sierra. Much of it has been destroyed by slash-burn agriculture as in the foreground. *Bursera simaruba* (L.) (right center) is one of the most characteristic trees.

Plate XV. Semi-open Forest-palm-grassland

1 km. north Tibernal, 100 m. a.s.l. On the southwestern slopes of the range at low elevations there is an extensive area which has been modified by burning, grazing and agricultural activity. Seral stages of forest remnants are mixed with palms, dense thickets, and grass.

Plate XVI. Pine-oak Forest

Ocotal Chico, 550 m. a.s.l. Forests of tall pines, *Pinus oocarpa* Schiede, in the southeastern massif are confined chiefly to ridges. The pines mix with oaks and occasionally other broadleaf forest trees on the slopes of ravines and small valleys. Volcán Santa Marta is visible in the right background.

in some places and in dense clumps in others. I saw little evidence of disturbance in this formation. Oaks are usually not mixed with the pines except at edges of the ravines where they sometimes occur in small groups. The dominant oak species is *Quercus ghiasbreghtii* Mart. & Gal., vel aff. Scattered forest remnants composed of the oak *Quercus peduncularis* Née, vel aff. grow on the slopes of volcanic cones northwest of Lago Catemaco. In addition to the grass and shrubs some large stands of the low shrub *Calliandra grandiflora* (L'Her.) Benth., vel aff. occur in the ground cover of the Volcán Santa Marta pine-oak formation and in secondary vegetation elsewhere.

The occurrence of pine forest on these southern slopes of the Sierra about 150 kilometers from the nearest pine area in Oaxaca raises the questions of how such a forest first became established here and why pines are not also present on the northwestern massif. *Pinus oocarpa*, the pine species in the Sierra, is widespread in Mexico, ranging from Sinaloa to Zacatecas and south through Chiapas into Guatemala (Standley, 1920:58). Budowski (1959:275) indicated that this species was one of four that can be found below 1000 meters elevation, but was more common above. It is very unlikely that wind transport of seeds could have bridged the long distance and intervening lowlands between the pine areas where they now exist. With the present topographic situation, it is not possible for them to have been water borne. The two most probable means of transport, if they originated this way, are by birds or human beings. Although it is not possible to rule out one or the other of these methods, transport by birds appears to me to be the most likely. The distance could easily be covered by a number of seed-eating species. Introduction by man could also have occurred at some time during the long settlement of the area, but the deliberate planting of pines would seem improbable since tree planting by people in the region has been known to involve exclusively broad leaved species mainly for fence rows and fruit yield.

The possibility also exists that the Sierra's pine area is a relict from a once continuous pine forest connection with that in the mountains of Oaxaca. Even though there is no geological evidence of a highland connecting the Sierra with the inland mountains, pines could have extended across the lowland gap between the inundations from which southern Veracruz finally emerged during the late Tertiary Period. The connection may have been severed by an inundation shallow enough to exclude the present pine area, or possibly the disruption was caused by a gradual change in edaphic conditions. Considering the areal and altitudinal adaptability of pine it seems unlikely that climate changes, which apparently were not pronounced in the region since the Tertiary, could have been a major cause of such a disruption.

The presence of pine forests in various regions has been the basis for diverse explanations. They are widely distributed on different soils, over widely differing rock formations and over a great range of altitudes. The broad ridge tops and their upper slopes on which most of the pines grow in the Santa Marta massif have a red clayey soil that appears to be old and well weathered. Surface layer acidity is high, more than that of the soil in most other parts of the Sierra. Friedlaender (1923:169) mentioned the "thick lateritic disintegrating older basalt lava of Santa Marta." Surface runoff of water is rapid and in many sections the soils are thin, thus promoting good external drainage. Crops do not produce well on this soil and this is the reason for a movement of people to better crop growing areas to the north. Although the Sierra's pines are on the drier side of the mountains, rainfall is high, probably ranging from about 1700 to possibly well over 3500 millimeters annually in the pine zone. Seifriz, in his study of Cuba's plant life (1943:404), believed that pines could not stand competition from other trees and therefore developed on the poorer soils. He remarked that its tolerance of dryness and preference for light serve it well on dry mountain ridges. In connection with the underlying basic rock on the pine areas of the Sierra, it is of interest to note that Seifriz (1943:417) found a large area of *Pinus cubensis* above 1000 meters on the south slope of Pico Turquino in the Sierra Maestra. He said the underlying rock, granite, determined the presence of pine here. Also, on the Sierra de Nipe in Cuba (Carabia, 1945) the pine forests grew well on a red soil rich in iron derived from serpentine rock, a basic metamorphic type of eruptive origin. Holdridge (1945:77) considered the pine forests of Hispaniola to be the result of repeated burning in which the hardwood species that normally grew up were destroyed and the pines survived. Some of the pines in the Sierra de Tuxtla showed old evidence of fire, but I saw no sign of extensive burning at the end of the dry season in 1962.

By whatever method the pines first reached the Sierra's ridges, I think that both edaphic and human factors probably are responsible for their persistence in this restricted area. The old and highly weathered

63

soil of these ridges provides favorable conditions for a pine forest to develop, especially where more open areas allow the heavy precipitation to accelerate weathering. Since there are now sections of broadleaf forest bordering the pines, especially on the ravine slopes and along the streams, it is probable that when better soil conditions existed such semideciduous forest occupied at least some of the ridges as the gum-oak forests above and the oak forests below the pine area do now.

Human settlement in the Sierra de Tuxtla dates from preconquest times, but no old village or ceremonial sites have been reported in the pine-oak section and I did not find evidence of any. Felling and burning of the broadleaf forest in this area to provide agricultural land provided more area suitable for pines because of decrease in soil fertility and reduction in vegetation competition. The villages in this area are mostly of the line type on the ridge crests and it is here where the most disturbance is likely to have occurred.

Finally, there has been no recent volcanism in the pine-oak area to provide volcanic materials containing fresh plant nutrients. Consequently, the soils have become less fertile and more suitable for pine-oak expansion than those which have been subject to comparatively recent ash falls on the Volcán San Martín Tuxtla massif. Some areas on the southern slopes of the latter massif do have less fertile soils and support small sections of oak forest. One possible explanation for the absence of pines there is that they never became established initially from the pine forest outside the Sierra.

Oak forest. There are extensive woodlands comprised almost entirely of oak on the lower slopes of the Volcán Santa Marta massif. The trees here are open-spaced with angular trunks and limbs, and average about 10 to 12 meters in height (Plate XVII). A short grass of variable density covers the ground with scattered shrubs or occasional thickets evident. In general this formation has not undergone major modification except where the grass and shrubs have been burned. Soil is often sandy and shallow so that rocks are exposed. *Quercus peduncularis* Née, vel aff. is a common oak tree species.

Gum-oak forest. Although *Liquidambar styraciflua* L. occurs sporadically in the humid forest on Volcán San Martín Tuxtla and exists in small stands on hill slopes northeast and northwest of Lago Catemaco, the largest areas of this formation are found on the southern slopes of

Plate XVII. Oak Forest

1 km. southeast Guasuntlan, 100 m. a.s.l. Open, low forest with grass ground cover shown here occupies a large area on the southeastern massif; the formation is occasionally mixed with savanna. A common oak species here is *Quercus peduncularis* Née, vel aff.

Plate XVIII. Gum-oak Forest

4.5 km. north Ocotal Chico, 850 m. a.s.l. The photograph shows the large, straight-trunked trees and the medium density of the sapling understory in mature forest. Many of the gum-oak forests in the range are secondary growth.

the Volcán Santa Marta massif. It has been much modified, however, and many seral communities exist. Trees in the mature forest are tall and straight trunked and there is an understory of medium density (Plate XVIII). One specimen on a ravine slope was 1.8 meters in diameter at breast height, but most larger trees are less than a meter in diameter. Almost pure stands of gum are occasionally intermixed with the oak, *Quercus ghiesbreghtii* and some other tree species. Gum-oak forest merges into the humid montane rain forest near the crest of the range. On apparently burned mountain ridges in this massif, the fern *Pteridium aquilinum* var. caudatum (L.) Sadela is common among dense, low shrubs and grasses. I judge that the gum-oak, pine-oak and oak forests on the southern slopes of the Volcán Santa Marta massif have expanded at the expense of tropical forest, which probably occupied much of the area, because of a combination of factors mentioned previously. These are vegetation destruction by man, and lowered fertility from long weathering, little addition of fresh volcanic materials and cropping. Another contributing factor may be the decreased precipitation in the rain shadow on these southern slopes. The broad, high front presented by the south crater walls of Cerro Campanario and Volcán Santa Marta is not duplicated in the northwestern massif, and thus a more effective orographic barrier exists.

Savanna. This formation frequently intermixes with oak forest or semideciduous forest at low elevations chiefly on the southeastern slopes of the mountains. It is an open formation consisting of small, angular trees, widely scattered, with a ground cover of variable density consisting of low to medium length coarse grasses, some sedges, and occasional shrubs. Prominent trees are *Curatella americana* L. (sandpaper tree) and *Byrsonima crassifolia* (L.) DC. (*nanche*); *Apeiba tibourbou* Ambl. also occurs in the savanna, in the mature semideciduous forest formation as well as in low, dense secondary forest. I would place this savanna in the "Orchard" phase of Beard's "Tall Bunch-Grass Type" (1953:190-192). The medium to large sized *Quercus oleoides* S. & C. is a fairly numerous tree where savanna and open semideciduous forest exist and in oak forest as well. I also found *Spondias mombin* L. (hog plum), a fairly large, sparsely branched tree, in the open type of semideciduous forest and in the savanna formation.

Savanna occupies small areas in the Sierra. It is mostly below 150 meters elevation. The land surface in this formation is gently

67

undulating and very gradually rises toward the north. Surface soil varies from grayish to dark brown, is friable, and appears to be a clay loam, becoming sandy in some sections. Exposed rocks, apparently basalt and tuff, occur frequently; the soil is thin in such places but elsewhere it seems fairly deep.

Total annual precipitation in the savanna areas averages about 1700 millimeters. Records at the nearest station, Guasuntlan, and at other stations in the foothills on the southern side of the range, show five months (January to May) and occasionally six (December) with less than 100 millimeters of precipitation. According to Beard (1953:196) and others, a month with below this amount can be considered a drought period in the moist tropics. It appears, therefore, that in this savanna there is a sufficiently long dry period to allow the ground vegetation to become combustible.

Although I did not observe the effects of a rain on this savanna, I expect that internal drainage would be poor considering the low relief in most parts of the formation. It is uncertain whether there is internal drainage impedance by a claypan or the bedrock. Considering the frequency of outcrops, however, bedrock may contribute to poor drainage.

These savanna areas are little used either for agriculture or grazing. I did observe some felling and burning of the semideciduous forest in the vicinity of the savanna, but saw no extensive burning and none in the savanna itself. The present population of these savanna areas is low; only a few small villages and occasional huts are within or adjacent to the formation. Budowski (1959:266) indicated that frequency of fires in the tropics was in direct relation with the density of population in regions where fires are possible. I think that extensive man-made fires in these savannas are infrequent or do not occur now. No large formerly inhabited sites have been discovered within the savanna areas, but there is reason to believe that the section was used during colonial times and probably also in the preconquest period. In view of this and the considerable areas of oak and semideciduous forests surrounding and within the savanna, it is possible that lightning-caused and man-made fires have occurred frequently enough in the past to have resulted in savanna expansion at the expense of the other forest formations. The survival of fire resistant trees (e.g., *Curatella*, *Byrsonima*) and the propensity of people in tropical areas to remove forest, points brought out by Budowski (1959:268 271) and others, could have been factors in the formation of these limited savanna areas. Budowski (1959:273) also said that all experiments conducted so far show that there is no savanna which eventually would not revert to forest where protected from fires and when seed source is nearby. A condition for such a process exists in parts of the savannas where oaks and trees from the semideciduous forest are intermixed with the savanna tree species.

The cause of the savanna here cannot be ascertained from present evidence. An intensive investigation might indicate that this savanna is the result of both fire in connection with the drought periods, and edaphic conditions, with the former being active in its early development, occasionally contributing to its expansion at present, and the latter aiding in its preservation.

Forest remnants, tree rows, thickets and open fields. Wherever the forest has been cleared and the land employed permanently for agriculture or grazing, vegetation in the form of tree rows, dense thickets and weedy fields is characteristic of the landscape. Most of such land lies in the more densely populated southern and southwestern parts of the Sierra, but as can be seen on Figure 7, there are large areas near Bahía Sontecomapan and Punta Roca Partida. Remnants of primary and secondary rain and semideciduous forest are found in the humid and drier sections respectively (Plate XIX). *Urera elata* (Sw.) Griseb. and other species of nettles are fairly common in the humid areas at forest edges, in openings and along trails, especially in secondary growth. Common trees growing from green fence posts placed in field edges are *Bursera simaruba*, *Gliricidia sepium* (Jacq.) Steud. (called *coquite* by the local people) and *Erythrina americana* Mill., in addition to those such as cultivated zapotes and mangos. Shrubs found abundantly in areas where successional vegetation was growing up are *Polymnia maculata* Cav., *Cordia spinescens* L., *Hamelia patens* Jacq., *Piper auritum* H.B.K., and *Conostegia xalapensis* (Bonpl.) DC. Botanists from the forestry research station in Mexico City listed species of *Zexmania*, *Desmodium* and *Iresine* as common in secondary growth near Zapoapan. Weedy fields are extensively overgrown with the low composite *Melampodium divaricatum* (Rich.) DC., its yellow flowers being conspicuous in late spring.

69

Plate XIX. Secondary Forest

3 km. east-northeast La Victoria, 150 m. a.s.l. This photograph shows a trail through humid forest in an area where much disturbance of the vegetation has occurred. Most of the trees are of medium to small size with few large ones remaining from the primary forest. There is a dense understory of shrubs, saplings, bamboo and the spiny palm *Astrocaryum mexicanum* Liebm. The large-leaved *Heliconia latispatha* Benth. occurs at edges and openings.

Littoral and mangrove. The volcanic headlands fronting the Gulf and the intervening stretches of shore possess a distinct type of vegetation. The headland faces support low, windswept trees, shrubs and herbaceous plants subject to salt spray. Some of these plants have sclerophyllous leaves or leaf parts and are evergreen. Along the ridge bordering the sand beaches grow strips of small, gnarled trees, occasional cacti, and dense thickets interspersed with coarse grassy areas, (Plate XX), reflecting the exposure to wind and salt spray as well as the lower annual precipitation. Littoral vegetation is variable in width, sometimes extending several hundred meters inland. Its representation on Figure 7 is necessarily not always to scale. Mangrove formation is restricted to the Bahía Sontecomapan where it attains sufficient development to form in places a closed forest up to 20 meters in height. *Rhizophora mangle* L. is a common species in this type.

Plate XX. Gulf Shore

Toro Prieto, sea level. Scattered cactus *Opuntia* sp. and other xerophytic plants occur on the narrow sandy ridge of the low coastal strips
between volcanic headlands. Punta Roca Partida is in the center, Isla
Terrón at left.

Chapter 3

FAUNAL ELEMENTS

3.1 Introduction

The size of the Sierra de Tuxtla and the large proportion that is essentially wilderness precluded a thorough zoogeographical study of its avian and mammalian faunas. My investigation of the birds and mammals, therefore, was limited to brief visits in various habitats throughout the region. I was able to secure information on the areal and altitudinal distribution and relative abundance of most nontransient bird species and on the twenty species of larger forest inhabiting mammals that I selected for study. During the field work of 1960 and 1962 I recorded 57 bird species that had not been reported previously from the Sierra. The nature of my investigation permitted ecological analysis of the birds and mammals principally on an areographical level. The length of my stay enabled me to observe almost a full cycle of seasonal activity, but much remains to be learned about the resident birds and mammals.

3.2 Life Zones

The Sierra de Tuxtla is altitudinally and areally within the Tropical Life Zone. Both the Humid Upper (Subtropical) and Lower Tropical Zones occur in the range, encompassing the extensive rain forest and the restricted areas of cloud forest. The Arid Division of the Lower Tropical Zone also extends irregularly into the Sierra on the inland side. This zone includes mainly the semi-open palm-forest-grass areas, the pine-oak, oak and tropical semideciduous forests and savanna. Its boundary with the Humid Lower Tropical Zone is transitional and tongues of the latter extend well into the Arid Zone in ravines and deep valleys. Loetscher (1941:63) gave the approximate upper limit of the (Lower) Tropical Zone as 2700 feet (823 meters) in Veracruz, and further divided this into upper and lower sections meeting in central Veracruz at 1200-1500 feet (365-457 meters) above sea level. Lowery and Dalquest (1951:537) considered 1700 feet (518 meters) as an approximate lower average elevation of the Upper Tropical (Subtropical) Life Zone. Goldman (1951:316) said that the Lower Tropical Zone in southern Mexico extends from sea level to about 2500 feet (762 meters) on north slopes and to about 3000 feet (914 meters) on south slopes.

My investigations show that there are no distinct life zone boundaries in the Tuxtlas. In regard to zonal indicators, Wetmore and Loetscher mentioned that bird species of the Upper Tropical Zone element range to lower altitudes, particularly during the months of the northern winter. I found this to be the case even during the warmer periods of the year when species such as the Purplish-backed Quail-Dove *(Geotrygon lawrencii)*, White-throated Robin *(Turdus assimilis)*, Slate-colored Solitaire *(Myadestes unicolor)*, Slate-throated Redstart *(Myioborus miniatus)*, Common Bush-Tanager *(Chlorospingus opthalmicus)* and Chestnut-capped Brush-Finch *(Atlapetes brunneinucha)*, occur frequently at 600 meters above sea level and even below 450 meters. Some mammals, for example, Howler Monkey *(Alouatta villosa)*, Spider Monkey *(Ateles geoffroyi)*, Tayra *(Eira barbara)* and Kinkajou *(Potos flavus)*, probably more characteristic of the Lower Tropical Zone in Mexico, range well upward into the higher zone. The Arid Division of the Lower Tropical Zone is also in most places not well defined, at least on the basis of the avifaunal elements. From an analysis of the temperature and moisture conditions and the distribution of plant formations it is possible to delineate the approximate life zone boundaries in these mountains, using as general indicators the altitudinal ranges of some characteristic bird species. Figure 8 depicts this life zone pattern. Although life zone definition is basically dependent on temperature in combination with moisture conditions,

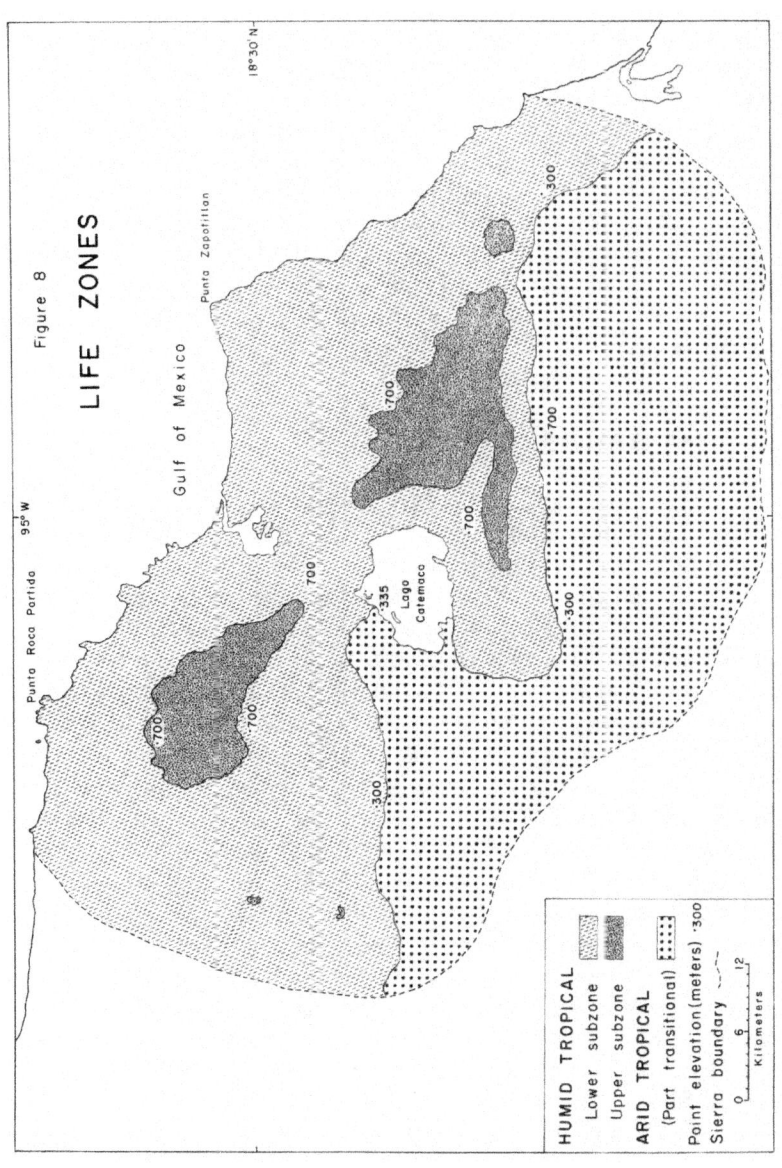

Figure 8

LIFE ZONES

HUMID TROPICAL
  Lower subzone
  Upper subzone
ARID TROPICAL
  (Part transitional)
Point elevation(meters) ·300
Sierra boundary ⌒

0   6   12
Kilometers

the former is not significant with respect to permanent environmental changes through the comparatively small relief of these mountains at this rather low latitude. Annual rainfall and the length and severity of the dry season are of more value in determination as is evident in the boundary location of the Arid Division, but are much less so in regard to the border between the humid zones. The transitional nature of these boundaries, particularly that between Arid and Humid Tropical, is evident from the altitudinal movement and mixing of the faunal elements. A detailed study of the Sierra's flora might enable a more definite demarcation to be made between the Humid Upper and Lower Tropical Zones.

3.3   Avifauna

Composition. A total of 384 species of birds representing 62 families have been recorded in the Sierra de Tuxtla region as outlined in this study. The abrupt descent of the seaward slopes to the Gulf in many places and the continuity of forest habitat there make sea level a more practical delimitation for avifaunal analysis of nontransients. There are comparatively few species restricted only to these low-lying areas. The 100 meter contour inland boundary marks a transitional level through which many species move, but serves to exclude in general most of those characteristic of true lowland avifauna.

There is no evidence of breeding for many birds that seem to be resident in the Sierra. On the basis of their frequency and time of occurrence, abundance and behavior, however, most of the species in question probably breed in the region. Nontransient species recorded number 263, representing 60 families. This group includes 39 species that are probably only visitants, several having been recorded only once or twice; since there is suitable breeding habitat in the region for them, it is possible that some actually do so at the present time. All of them breed in other parts of Mexico. A few species are resident in the Sierra only during part of the year, migrating southward for the northern winter months. Appendix C contains a complete list (partially annotated) of the nontransient bird forms that, to my knowledge, have been recorded in the region.

Table II shows that in the composition of the nontransient avifauna

76

TABLE II			
NUMBER OF NONTRANSIENT BIRD SPECIES BY FAMILIES RECORDED IN THE SIERRA DE TUXTLA			
Family	Breeds	Probably Breeds	Visitant
Tinamidae (tinamous)	4		
Podicipedidae (grebes)	2		
Pelecanidae (pelicans)	1		
Phalacrocoracidae (cormorants)			1
Anhingidae (anhingas)			1
Fregatidae (frigate birds)	1		
Ardeidae (herons, bitterns)	4	2	4
Cochleariidae (Boat-billed Herons)		1	
Ciconiidae (storks)			1
Threskiornithidae (ibises)			1
Anatidae (ducks, geese, swans)			2
Cathartidae (New World vultures)	2		1
Accipitridae (hawks, eagles)	6	1	6
Pandionidae (ospreys)			1
Falconidae (falcons, caracaras)	3	2	1
Cracidae (guans, curassows. chachalacas)	3		
Phasianidae (partridges, quails)	2		
Aramidae (limpkins)		1	
Rallidae (rails, gallinules)	2	3	1
Heliornithidae (sungrebes)	1		
Jacanidae (jacanas)	1		
Charadriidae (plovers)			2
Laridae (gulls, terns)			4
Columbidae (pigeons, doves)	10	1	2
Psittacidae (parrots)	5		
Cuculidae (cuckoos, anis)	3	2	
Tytonidae (barn owls)		1	
Strigidae (owls)	3	2	
Nyctybiidae (potoos)	1		
Caprimulgidae (nightjars)	1		1

Family	Breeds	Probably Breeds	Visitant
Apodidae (swifts)	2		1
Trochilidae (hummingbirds)	14		
Trogonidae (trogons)	4		
Alcedinidae (kingfishers)	4		
Momotidae (motmots)	2		
Galbulidae (jacamars)	1		
Ramphastidae (toucans)	3		
Picidae (woodpeckers)	9		
Dendrocolaptidae (woodcreepers)	6		
Furnariidae (ovenbirds)	4		
Formicariidae (antbirds)	3	1	
Pipridae (manikins)	1		
Cotingidae (cotingas)	5		1
Tyrannidae (tyrant flycatchers)	22	2	2
Hirundinidae (swallows)	3		
Corvidae (crows, jays)	2		
Paridae (titmice)		1	
Troglodytidae (wrens)	5		
Mimidae (mockingbirds, thrashers)			1
Turdidae (thrushes)	5		1
Sylviidae (gnatcatchers, kinglets)	2		
Cyclarhidae (peppershrikes)	1		
Vireolaniidae (shrike-vireos)	1		
Vireonidae (vireos)	3		
Coerebidae (honeycreepers)	2		
Parulidae (wood warblers)	6		
Ploceidae (weaver finches)	1		
Icteridae (blackbirds, orioles)	11		3
Thraupidae (tanagers)	13		1
Fringillidae (finches, grosbeaks, sparrows)	14		

about half of its species belong to only about 10 per cent of the families; the Tyrannidae possess by far the largest number of members (26) with the Ardeidae, Accipitridae, Columbidae, Trochilidae, Icteridae, Thraupidae and Fringillidae having at least ten each. Most families have less than six members each and one-third are represented by only one species each.

The Tinamidae and Alcedinidae are well represented in the Sierra, each with four of the five species resident in Middle America. Other families with a relatively high percentage (30 or more), excluding those with only one or two members in Middle America, are the Ardeidae (60 per cent), Cathartidae (75 per cent), Accipitridae (39 per cent), Falconidae (46 per cent), Cracidae (37 per cent), Columbidae (43 per cent), Ramphastidae (37 per cent), and Dendrocolaptidae (30 per cent). The large families Trochilidae and Fringillidae are relatively poor in numbers of species (14 per cent), and the Phasianidae, Pipridae, Corvidae and Mimidae have less than 1 per cent of their Middle American forms occurring as nontransients in the region.

Habitat. The fairly great variety of habitats in the Sierra de Tuxtla and the fact that many species are not restricted to one type make it difficult to assign some species to particular habitats. Appendix C has six major habitats based mostly on distinctive vegetation associations, and contains the species that are primarily associated with each. After each species is a general indication of its average abundance in the Sierra in relation to other species in the habitat. There is often variability in numbers at different seasons. I based abundance on actual numbers and frequency of occurrence from my own records and those of other investigators.

In Appendix C the essentially nontransient avifauna of 262 species, excluding the House Sparrow *(Passer domesticus)*, is distributed in the various habitats as follows: 1) habitat A - 95 species characteristic of tropical rain forest including cloud forest, 2) habitat B - 81 species which occur principally in the forest edge, thickets, bush and tree rows and fields in humid areas, 3) habitat C - 20 species characteristic of semideciduous tropical forest, forest edge, thickets and bush and tree rows in drier areas, 4) habitat D - 4 species characteristic of pine and oak forest, 5) habitat E - 15 species which range widely in humid and dry open areas, and 6) habitat F - 47 species which are usually

79

associated with water bodies in the form of sea, lakes, swamps and streams.

Areal distribution and relative abundance. Of the 95 species in habitat A, 20 (21 per cent) also frequent different types of less humid habitat; in contrast, however, 61 (75 per cent) of the 81 species in habitat B range into drier areas. The low percentage of forms in habitat A occurring in drier areas is an indication of the rather restricted ecological requirements of the tropical rain forest species. Twenty-eight families are represented in the rain forest habitat type. The Tinamidae, Cracidae, Phasianidae, Psittacidae, Trogonidae, Momotidae, Galbulidae, Ramphastidae, Dendrocolaptidae, Furnariidae, Pipridae, Turdidae, Vireolaniidae, and Vireonidae have all their members or all but one characteristic of the habitat. The Fringillidae and Icteridae, however, have only about 3 and 1 per cent respectively represented. In considering relative observable abundance, a comparative analysis of the species' status in habitats A and B, the two comprising most of the areal extent of the Sierra, shows about 50 per cent of the species in the latter habitat type as fairly common to abundant, but only 35 per cent of those in the former type are classed as such. Thus, despite the greater number of species in rain forest habitat, they are generally less abundant than are those in the other major habitat. It is questionable whether some species are more characteristic of one habitat than another in the region, at least in relation to abundance. For example, the Streak-headed Woodcreeper *(Lepidocolaptes souleyetii)*, though occurring in rain forest regularly, also can be frequently observed in tropical deciduous forest and oak forest. The Keel-billed Toucan *(Ramphastos sulfuratus)*, Golden-olive Woodpecker *(Piculus rubiginosus)* and Bright-rumped Attila *(Attila spadiceus)* are other examples. In such cases I have tried to assign the species to the habitat type in which it appears to be most numerous.

The majority of the Cuculidae, Cotingidae, Corvidae, Icteridae and Fringillidae are found in habitat B. Almost half of the large Tyrannidae family and the Thraupidae also occur in this type. Often some habitat B species enter the forest or frequent natural or man-made openings, but primarily they are birds of edge and more open sections. Although in general they are more abundant than the rain forest habitat species, many forms are irregularly distributed over the region; several, such

80

as the Red-eyed Cowbird *(Tangavius aeneus)* and the White-collared Seedeater *(Sporophila torqueola)*, concentrate in flocks and may be absent in many suitable areas. The distribution of the species within the humid forest is more regular, although small flocks of several species frequently occur, especially when not breeding. This even density of birds is to be expected in the relatively homogeneous habitat of the humid forest.

The forms characteristic of habitat C, besides being considerably less in number of species than those of the previous two habitat types, are markedly less abundant: only two of the 20 species are fairly common in the Sierra. Over half occur occasionally in the more humid parts of the range. Only two characteristic species are common in the D or pine-oak habitat, the Red-billed Azure-crown *(Amazilia cyanocephala)* and the Acorn Woodpecker *(Melanerpes formicivorus)*. The former ranges up to the open ridge areas near the gum-oak forests and the latter also occurs at low elevations in open semideciduous tropical forest. Nine of the 15 widely ranging species are infrequently observed in the Sierra. The White-collared Swift *(Streptoprocne zonaris)* and particularly Vaux's Swift *(Chaetura vauxi)* often congregate in large flocks. Many species associated with or dependent upon water range widely in both humid and drier sections, inhabiting stream and lake shores as well as adjacent open fields and forest edges. There are few small areas besides Bahía Sontecomapan capable of supporting marsh birds. The Brown Pelican *(Pelecanus occidentalis)* and Magnificent Frigatebird *(Fregata magnificens)* range into the Sierra and apparently breed on Isla Terrón in the Gulf near Punta Roca Partida. Bahía Sontecomapan supports a number of the Ardeidae, most also occurring at higher elevations. The Sungrebe *(Heliornis fulica)* and Gray-necked Wood-Rail *(Aramides cajanea)* have been recorded near sea level, but the latter may occur at higher levels. The members of the Charadriidae and Laridae have been observed at sea level and in the vicinity of Lago Catemaco and could breed on the more remote coastal areas. The most numerous species in habitat type F are the Olivaceous Cormorant *(Phalacrocorax olivaceus)* and the Snowy Egret *(Leucophoyx thula)*, both abundant on Lago Catemaco, but uncommon elsewhere.

Altitudinal distribution. Altitudinal distribution of birds correlates not only with climatic elements as evidenced by the different vegetation formations, but with the latter's modification by the human element. The maximum altitude of the Sierra is low (1660 meters) compared to other mountain ranges in Mexico, and vertical temperature change (normal lapse rate) is not great. Normal altitudinal ranges of most birds, therefore, do not seem to be affected by temperature. A few species' vertical ranges extend from sea level to the highest points of the Sierra. Yet most occur within smaller altitudinal limits and the distribution of the various original plant formations and their modifications seems to be the principal governing factor. Therefore, in Appendix C I have included the approximate altitudinal ranges above sea level for most species. These limits are in some instances tentative and it is likely that the upper altitudinal limits of a number of species, such as the Black-chinned Jacamar *(Galbula ruficauda)* and Scaled Antpitta *(Grallaria guatimalensis)*, are greater than is indicated. For species recorded once or twice specific elevations are given.

The species' elevations in the humid forest habitat type show that about two-thirds of the forms were not recorded above about 950 meters elevation. The one-third observed above this elevation are those species ranging over most of the altitudinal extent of the Sierra and those typical of the Upper Tropical Zone. That some of the latter occur commonly to much below the 950 meter level, especially during less favorable weather from October to March, indicates the ill-defined nature of the Upper Tropical Zone here. As Wetmore (1943:223) pointed out, such species tend to descend lower here than in the mountains of Central America, being affected by the lower temperatures brought by heavy rains and the colder storms from the north.

The great majority of bird species in the region are primarily residents of the Lower Tropical Zone. With the few exceptions of wide ranging species such as the Black Vulture *(Coragyps atratus)* and those frequenting all elevations, e.g., Red-billed Pigeon *(Columba flavirostris)*, Bar-tailed Trogon *(Trogon collaris)*, and Golden-olive Woodpecker *(Piculus rubiginosus)*, most range well below 1000 meters elevation. Only about 10 per cent of the Sierra's land is above 750 meters, an areal factor restricting the comparative number of birds at higher altitudes, and the continuous forest habitat there also limits the variety of species.

The interdigitation of the drier areas in the region by more humid sections of forest, especially in stream valleys and ravines, and the average annual rainfall even in the less humid areas being fairly high (usually over 1500 millimeters) with no prolonged dry season, tend to make the drier subzone less well defined. Therefore, I have not listed birds characteristic of the Arid Division of the Tropical Zone in the region. Those species contained in Appendix C, habitats C and D are most indicative of drier conditions.

History, endemism, affinities. Although Staehelin has indicated that the basement ridge of the Sierra de Tuxtla acted as a barrier against the sea advance during the middle Mesozoic, it is not certain whether it was continually emergent during the Tertiary inundations. The history of its present avifauna probably begins in the late Tertiary Period subsequent to the range's permanent emergence and greatest uplift in the Pliocene Epoch. As in other parts of Mexico and Central America, the Sierra was affected by the faunal mixing that occurred during the Tertiary and which probably increased after the closing of the water gaps between North and South America and the cooling of the climate in North America in the late Tertiary and Pleistocene. Thus, the Sierra has been open to avifaunal colonization from both north and south for only a comparatively short period of geologic time.

Even during the lowered temperature of the Pleistocene glacial stages it is unlikely that the climate of the Sierra ever became cooler than warm temperate. From the Eocene to the Pleistocene and in the latter's interglacial stages, it was probably tropical (Dorf, 1959:185-199). Thus, though tropical vegetation elements were less abundant, and the Upper Tropical Life Zone probably descended in elevation during the Pleistocene and perhaps earlier (Griscom, 1950:358), bird life was not impoverished or completely displaced. No doubt some bird species of tropical affinity moved southward and those of the Upper Tropical Zone to lower altitudes. When the climate ameliorated after the final glacial stage, the humid tropical vegetation elements proliferated and were able to attain their present abundance farther north. Bird species characteristic of the Upper Tropical Zone then ascended to higher elevations and established themselves, and the northern extension of tropical vegetation provided a continuous route for dispersal of Lower Tropical Zone birds from the Central American lowlands.

It is possible that there was some interchange of Upper Tropical Zone forms between the Sierra and inland mountains during the lowered elevation of this zone, but there is no geological evidence that the Sierra was ever connected to other ranges by a highland. Therefore, the physical isolation of the Sierra by lowlands has been of sufficient duration to enable endemic birds and other vertebrate forms to develop. Wetmore (1943:225) listed five Upper Tropical Zone endemic birds from Cerro Tuxtla and Volcán San Martín Tuxtla, and Lowery and Newman (1949:8) later determined another from those collected by Wetmore. The new endemic species of toad and tree frog described by Firschein (1950:84) and Firschein and Smith (1956:17) also were from Volcán San Martín Tuxtla. The latter paper mentioned that at least eight endemic forms of reptiles and amphibians have been reported from the range. All the endemics are considered to be fragments of the fauna represented during cooler conditions of the Pleistocene and preserved in isolation in the Sierra. I did not find any new endemic forms in the region. There remains the possibility of discovering endemics, however, if intensive study is undertaken at the high elevations, particularly in the lesser known southeastern massif.

I found five of the six endemic birds to be well distributed in the humid forest throughout the mountain range mostly at higher elevations. The minimum altitudinal limits of four, however, vary from only about 300 to 600 meters above sea level, the fifth not ranging below about 900 meters and the sixth, the Wedge-tailed Sabrewing *(Campylopterus curvipennis excellens)*, occurring down to sea level. Despite the relatively low minimum altitudes attained by several of these birds, the subspecific differentiation of all except the last mentioned was assured by ecological isolation in two ways. First, they are adapted to conditions of high humidity, and to cooler temperatures than those extant in the lowlands. Second, they are restricted to a habitat of montane rain forest in the Sierra. Since forest types in the lowlands surrounding the Sierra de Tuxtla are different both in structure and plant composition from this montane forest, they do not present suitable habitat for the endemic bird forms, and thus serve as an effective isolating barrier. The extensive lowlands bordering the Sierra are markedly drier than the montane forest except where sections of swamp forest occur. An exception to this dryness is on the south-southeast where higher rainfall and more humid conditions extend from the Coatzacoalcos basin. The essentially less humid conditions, however, if they have persisted for a considerable length of time as seems likely, present a

84

significant element conducive to isolation of the Sierra de Tuxtla. The endemic hummingbird *Campylopterus curvipennis excellens* has been found outside the Sierra near Jesus Carranza (Lowery and Dalquest, 1951:583). It occurs to sea level in the Sierra and ranges readily into forest edge. Although conditions suitable for its primary differentiation exist at higher elevations in the Sierra, this hummingbird's mobility and adaptability enabled it to extend into the lowlands, particularly where more humid conditions exist, as in the Coatzacoalcos basin.

It is probable that additional endemic forms of more sedentary animals, such as amphibians, exist in the Sierra at upper elevations where relatively stable conditions have long persisted. The mobility and adaptability of birds probably precludes any except the ecologically restricted Upper Tropical Zone species from evolving into endemic forms in the Sierra. It is also possible that small endemic mammals may yet be found in the Sierra de Tuxtla, but from the presently known taxonomic status, range extent, ecological requirements and relative lack of plasticity of most larger tropical mammals, it appears unlikely that new forms of the latter will be discovered in the region.

Wetmore remarked on the survival of Upper Tropical Zone endemic birds on Volcán San Martín Tuxtla despite the recent volcanic activity there. The three eruptions in the first third of the 16th century, 1664 and 1793, which apparently created only local alterations of habitat, are the sole disturbances on record for the Sierra. Some of the volcanic cones and ridges of the Volcán San Martín Tuxtla and many of those in the Volcán Santa Marta massif appear to be at least several centuries old and not formed by any recent activity. The majority of the Upper Tropical Zone birds, including the endemic forms, also inhabit the Santa Marta massif; thus, the largely forested ridge north of Lago Catemaco forms a connecting route for an interchange of these species between the massifs, especially since a number descend to lower elevations during part of the year. Therefore, since it appears that volcanic disturbances after the Sierra's formation have not occurred simultaneously throughout the range, this route has possibly aided in the survival of the endemic species.

Although the Sierra has been open to bird movement from the north, its climate and most of its habitats are essentially tropical; nevertheless, it is sufficiently near the limits of the Neotropical region to permit forms of northern affinity in its avifauna. The nontransient bird species of the Sierra can be placed within the following elements (cf. Mayr, 1946): a) widely ranging families distributed in both tropical and temperate areas of the world and of obscure origin, b) Pantropical families widespread within the Old and New World tropics but of uncertain origin, c) Old World families originating in the Old World and arriving in the New from very early to recent times, some of the early arrivals establishing secondary evolutionary centers in America, d) North American families developed in North America during the Tertiary when the continent was separated from South America, e) Panamerican families being either originally North American or South American but having endemic genera and species in both, and f) South American families well developed in South America and scarce in Central America.

Table III gives the numbers of families and nontransient species representing each element in the Sierra de Tuxtla. These figures show that most families in the Sierra are included in the widely distributed element and this element is second only to the Panamerican in number of species. Prominent in the widely distributed element are the Ardeidae, Accipitridae, Falconidae and Picidae. It also includes the grebes, pelicans, ducks, gulls, shorebirds and rails. Elements of North American and Old World origin, many having arrived in the region from the north, total 21 families with 68 species, in contrast with the essentially southern origin of the Panamerican and South American elements, represented here by only 16 families, but including 111 species. The North American and Old World elements consist of such families as the Columbidae, Cuculidae, Strigidae, Turdidae, Troglodytidae, Parulidae and Fringillidae (subfamily Carduelinae). Many forms in the major families of the Panamerican element (Tyrannidae, Trochilidae, Icteridae) have moved far into North America, with the exception of those in the Thraupidae. Only a small percentage of some large South American element families (Furnariidae, Formicariidae) reach the region. Almost half of the species comprising these last two elements are characteristic of the humid forest forming a continuous route for their immigration from the south. The avifauna of the Sierra, therefore, is composed of a large number of species in the widely distributed and Pantropical elements of unknown or uncertain origin (30 per cent of the nontransients recorded). Apart from these elements, species of southern origin predominate (42 per cent) in comparison to those with

TABLE III		
AVIFAUNAL ELEMENTS IN THE SIERRA DE TUXTLA		
ELEMENT	FAMILIES	SPECIES
Widely distributed	17	66
Pantropical	7*	14
Old World	11*	43
North American	11* **	25
Panamerican	5***	71***
South American	12* **	43
* Family Fringillidae included		
** Family Sylviidae included		
*** Includes one family and three species of northern origin		

more northern affinities (26 per cent), reflecting the essentially tropical nature of the more sedentary portion of the avifauna.

3.4 Mammalian Fauna

There has been little study of mammals in the Sierra de Tuxtla. Nelson and Goldman did some work on mammals during their brief stay in 1894, and Leopold visited the region while engaged in his research on game animals. The geographic ranges in Mexico of 94 mammal species include the Sierra de Tuxtla (Hall and Kelson, 1959). Table IV gives a list of the mammal families and the number of species in each occurring or possibly occurring in the region. The Two-toed Anteater *(Cyclopes didactylus)* and the Hooded Skunk *(Mephitis macroura)* also may range into the Sierra. Almost 50 per cent of the mammals are bats. I selected for study 20 species of mammals from ten families on the basis of a) habitat (most are primarily inhabitants of humid tropical forest, the major remaining undisturbed plant formation in the region), b) size (they and their signs are relatively easy to observe and they are known by many of the local people), c) abundance (several are rare and others are decreasing due to various causes), d) food source (a number are often utilized by the local people for subsistence). Although part of the information I secured on these species is from my own observations, a considerable amount is necessarily derived from reports of inhabitants, particularly hunters, from all sections of the Sierra. Obviously, this latter source varies in reliability and accuracy, a factor that must be considered at all times in evaluating such information.

Habitat, distribution and abundance. All but four, the River Otter *(Lutra canadensis)*, Jaguarundi *(Felis yagouaroundi)*, Mountain Lion *(Felis concolor)* and Collared Peccary *(Pecari tajacu)*, of the twenty species of mammals discussed here are primarily inhabitants of humid tropical forest in the Sierra and dependent upon this habitat for survival. Several may occasionally range into the drier areas of semideciduous forest, but these are not extensive in the region. Consequently, the distribution of the mammal species considered here corresponds approximately to the area of unbroken humid forest in the range, except for the four species mentioned above; *Felis yagouaroundi* and *Pecari tajacu* often frequent forest edge or dense thickets beyond the forest proper. *Felis concolor* ranges in various habitats and *Lutra canadensis* occurs entirely in and near water. An account of the status of each species follows.

The Howler Monkey *(Alouatta villosa)* is a fairly common resident in the Sierra. It occurs from near sea level to at least 1300 meters in the primary rain forest and cloud forest. I recorded no more than two of these monkeys together and one hunter had seen no more than four in a group, but Pedro Mateo of Piedra Labrada reports 15 or 20 seen at one time. They may be more numerous on the Gulf slopes of the Santa Marta massif than they are north of Volcán San Martín Tuxtla. I recorded from one to five individuals on the latter volcano and on Volcán Santa Marta, near Cerro Campanario, above Colonia Huatusco, near Montepío, above Sontecomapan and near the Cumbres de Bastonal and the Río Carizal. I do not believe that Howler Monkeys are hunted to a great extent, although no doubt they are killed when there is opportunity. Some people of Ocotal Chico believe that there are as many monkeys as there have ever been. Locally, this may be true, but the continual removal of the tall rain forest has certainly decreased their total number. The Howler is commonly called *mono zambo*, sometimes *aullador* and rarely *saraguato*. It is known as *bumbumutsu* by the Popolucas of the Ocotal area.

TABLE IV	
MAMMAL FAMILIES AND NUMBER OF SPECIES IN EACH WHICH OCCUR OR MAY OCCUR IN THE SIERRA DE TUXTLA*	
Didelphidae (opossums)	4
Soricidae (shrews)	2
Emballonuridae (sac-winged bats)	4
Phyllostomidae (American leaf-nosed bats)	22
Desmodontidae (vampire bats)	2
Natalidae (funnel-eared bats)	1
Vespertilionidae (vespertilionid bats)	9
Molossidae (free-tailed bats)	7
Cebidae (monkeys)	2
Myrmecophagidae (anteaters)	1
Dasypodidae (armadillos)	1
Leporidae (rabbits)	2
Sciuridae (squirrels)	2
Geomyidae (pocket gophers)	1
Heteromyidae (heteromyids)	2
Cricetidae (cricetids)	8
Erethizontidae (New World porcupines)	1
Dasyproctidae (agoutis, pacas)	2
Canidae (wolves, coyote, foxes)	2
Procyonidae (raccoons)	4
Mustelidae (mustelids)	5
Felidae (cats)	5
Tapiridae (tapirs)	1
Tayassuidae (peccaries)	2
Cervidae (cervids)	2
* Hall and Kelson (1959)	

The Spider Monkey *(Ateles geoffroyi)* is called *mono chango* by the local people. It is a common mammal in many sections of the Sierra. In both primary and secondary forest it ranges from sea level to at least 1500 meters and also is sometimes found in the less humid forests. From two or three to about 20 or 30 is the usual variation in group size in the Sierra. I believe the Spider Monkey is subject to greater hunting pressure than is the Howler; occasionally young ones are captured and sold as pets.

The Tamandua *(Tamandua tetradactyla)* or *brazo fuerte* is generally uncommon through the Sierra. This anteater ranges into the upper elevations to at least 1000 meters. Almost everyone knew it, most indicating that there were a few in their respective areas. It was reported to be fairly common near Montepío, Ocotal Chico and the Bastonal. Gary Ross observed one at about 1000 meters elevation on Volcán Santa Marta. I found a dead anteater near the northwest side of Lago Catemaco; and a live one captured near the Río Cuetzalapan was brought to me.

I was not successful in finding the Forest Rabbit *(Sylvilagus brasiliensis)*. Apparently it is uncommon in this region as few of the local people knew it. Alfonzo Lazaro called it rare, but has found it in the forest above Los Mangos. Epigmenio Tegoma also knows it, but said there were very few near Tapalapan. José Arias, who lives near the mouth of the Río Carizal, had seen several in the nearby forest. A few were reported to be in the forests above Rancho Ahuacapan near Cuetzalapan. One of the Popoluca boys from Ocotal Chico stated that this rabbit came to eat the beans in his field on one of the south ridges of Volcán Santa Marta.

The *tepescuintli* or Paca *(Agouti paca)* is a fairly common species in the wilder sections of the Sierra. Its tracks and runways are frequently seen in the humid forest. As in other places in Mexico, its meat is highly prized by the people and the animal is hunted frequently. Paca meat is sold in the market in San Andrés Tuxtla. Although most people considered the Paca to be fairly common, several indicated that they have decreased markedly in numbers, especially near settlements. A live one trapped on the Río Yougualtajapan was offered for sale. In the elfin forest at 1640 meters on the north side of Volcán San Martín Tuxtla I met four hunters who had just shot a tepescuintli. Near Laguna Tisatal above Tapalapan Epigmenio Tegoma filled several deep

90

holes in a rock cliff, because Pacas took refuge in such places when pursued during hunts. Leopold (1959:388) found it "fairly plentiful" in the Sierra.

In many sections the Agouti *(Dasyprocta punctata)* is as numerous as the Paca and in some localities probably more so. Local people regard it as common. It is partially diurnal. I saw several during the day. Their trails and burrows were evident in the forest. It also is hunted frequently for meat; reports from some sections such as Ocotal Chico and Los Mangos indicate there are few or none because of the hunting pressure. Near Tapalapan they feed on the sugar cane in either day or night. I was told that there was a bounty of ten pesos each on them in this area.

I did not locate any Cacomistles *(Bassariscus sumichrasti)* in the Sierra. Most people were unacquainted with the species. Epigmenio Tegoma at Tapalapan and a man on the Río Yougualtajapan seemed to know the animal, but were not clear as to its status.

The *tejon* or Coati *(Nasua narica)* is fairly common in the Sierra. The largest number of individuals I observed in a group was ten southwest of Sontecomapan; bands of 40 to 50 are said to be not unusual in the forests above Piedra Labrada. They range high into the mountains. At 800 meters on the trail below Cerro Campanario I saw a hunter carrying a large Coati, probably a male. Near Laguna Tisatal, Epigmenio Tegoma fired at, but missed, a large solitary individual which ran down the trail ahead.

The Kinkajou *(Potos flavus)* is well distributed in the humid forest through the mountains. All the local people knew it well and called it *marta*. I observed them on Volcán San Martín Tuxtla, above Colonia Huatusco, on the Río Carizal and elsewhere. They seem to be entirely nocturnal and occur sometimes singly, but most often in pairs or a group of three. At night it is easy to detect them in the trees as they move noisily about feeding on fruit. Jeotulio Gutierrez said that they eat the coffee berries in the groves above Ocotal Chico. I occasionally saw Kinkajou skins in the villages.

The Tayra *(Eira barbara)* is generally distributed, but rather uncommon, in the region. I saw this large weasel only above Sontecomapan and high on Volcán San Martín Tuxtla. It is called "sugar cane cat" by some of the people near Piedra Labrada where they say it feeds on the cane. Most commonly it is known as *cabeza de viejo*. The Tayra

91

is reported occasionally to run down Red Brocket and is supposed to drink the blood and leave the meat. This weasel is also said to kill snakes including venomous species. One day at dawn at 1200 meters on Volcán San Martín Tuxtla I quietly watched three Tayras move past about three meters distant. They were extremely agile and when one perceived me it jumped up and down on a log snarling and hissing in an aggressive manner before making off into the forest.

No one I questioned knew the Grison *(Galictus allamandi)* in the region and I did not observe it.

The River Otter or *perro de agua (Lutra canadensis)* is known by most people living along the larger streams and is most numerous at lower elevations on the Gulf side (Figure 9). Estimates of numbers varied from two or three on a particular stream to 20 or 30 on the Ríos Col and Máquina above Montepío (Plate XXI). I believe the last figures are too high. The women washing clothes in the rivers in various places saw them in the daytime. José Arias said one or two came frequently close to his house on the Río Carizal. I did not see any during my field work. In addition to those rivers mentioned I was told that there were others on the Ríos Salinas, (Arroyo) del Oro, Palma, Yougualtajapan, Coxcoapan, Salado, Mescalapan, Tecuanapan and Sochiapan. Andres Baxin saw one or two in the Río Tular, a small stream north of San Andrés Tuxtla. There were also reports that the animal was rare in the Ríos Guasuntlan, (Arroyo) Hueyapan and Tecolapan on the southern and southwestern sides of the range. The River Otter does not seem to be hunted much in the region ("Santiago Tuxtla, Rincon de Maravilla," 1961:13).

According to Raul Argudin the Jaguar *(Felis onca)* is not uncommon in the Sierra de Tuxtla. He gave a cautious estimate of "about fifty." I saw an old skin in Catemaco from near Cuetzalapan. My other informants indicated that there were few and that the *tigre* is rare. A young hunter in the virgin rain forest above Colonia Huatusco said a cow had been killed by a Jaguar near there in early 1962. Andres Baxin spoke of a Jaguar killed about mid-May of the same year high on the southeast slope of Volcán San Martín Tuxtla. One was reported killed above Colonia Huatusco in 1961. I saw the track of a Jaguar at about 1100 meters on the southwest side of Volcán San Martín Tuxtla in 1962, and E. P. Edwards (personal communication) observed what he thought to be a track of this cat on Volcán San Martín Pajapan.

92

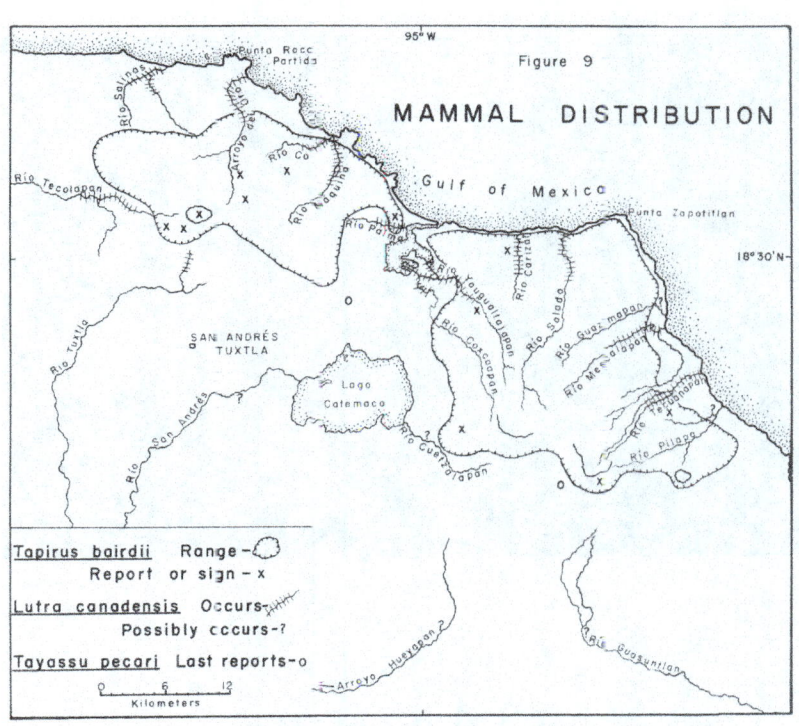

Figure 9

MAMMAL DISTRIBUTION

Gulf of Mexico

Tapirus bairdii   Range-
           Report or sign - x
Lutra canadensis Occurs-
           Possibly occurs-?
Tayassu pecari Last reports-o

0   6   12
    Kilometers

Plate XXI. Coastal Stream

1 km. south Montepío, 10 m. a.s.l. Low swamp or flood plain forest borders the Sierra's larger streams when they near the Gulf of Mexico. On the Río Máquina, shown here, the forest floor is occasionally swept clear of humus and leaves by inundations. The River Otter *(Lutra canadensis)* inhabits such waterways.

The Jaguar is apparently not often hunted in the region unless cattle are killed. From all indications the species is very uncommon and keeps mostly to the extensive forests on the Gulf slope of the range.

The Mountain Lion *(Felis concolor)* is rare in the Sierra de Tuxtla and most reports of it came from the southeastern part of the range. Jeotulio Gutierrez said that, unlike the Jaguar, it seemed to kill pigs rather than cattle or horses. Raul Argudin told me that he knew of three *leones* killed in about twenty years. In late May 1962, I noticed a fresh skin, measuring about six feet in length, hanging in the market in San Andrés Tuxtla. The animal was killed near Zapoapan in the vicinity of Cerro Cintepec about two or three weeks previously.

The *ocelote* or *tigrillo 'Felis pardalis)* is fairly numerous in the humid forests of the Sierra. I saw their tracks in a number of localities and their skins in various villages. Pedro Mateo said that this cat is a nuisance catching chickens around the new settlements near Piedra Labrada. Andres Baxin shot one on Volcán San Martín Tuxtla in November 1962, and one was killed on the Río Yougualtajapan in August. Jeotulio Gutierrez said that they are often killed above Ocotal Chico. A young tigrillo for sale had been captured northwest of Lago Catemaco. The people did not distinguish the Margay *(Felis wiedii)* and I did not record it.

The *onza* or Jaguarundi *(Felis yagouaroundi)* is a fairly common resident in certain sections and prefers thickets and forest edge to the unbroken forest. Because of this habitat preference it is more conspicuous than the other cats. Several people remarked that it came close to the villages and often caught domestic chickens. I observed a red phase onza crossing the road about two miles north of Santiago Tuxtla and noted another in the same color phase near Laguna Tisatal. R.W. Dickerman saw a Jaguarundi at about 600 meters elevation on Cerro Cintepec.

A report of game birds and mammals in the region compiled by Elias González, Jefe Sector Forestal at San Andrés Tuxtla (1962), notes after the tapir, *"casi agotado"* (almost exhausted). Raul Argudin estimated that "about 100" remain. I was given estimates of about 50 above Montepío and approximately 30 on and about Volcán San Martín Tuxtla. There is no doubt that the Baird's Tapir *(Tapirus bairdii)*, called *danta* or *antsburro*, has decreased considerably in numbers in the past few decades; it is certain, however, that a small number remain in the unbroken humid forests and stream valleys of both massifs.

Raul Argudin knew of about seven tapirs killed in the last 20 years, including two shot near La Palma in 1961, on the west side of Bahía Sontecomapan. In his book he recounted (1955:33) the taking of several, including one of "300 kilos." I heard a report in San Andrés Tuxtla that a hunting party had killed three, one a young animal, in February 1962, between Montepío and Volcán San Martín Tuxtla. José Arias shot an adult and young in the forest of the Río Carizal valley. Two or three are killed annually near Montepío.

Above Laguna Tisatal in June fresh tapir tracks had been seen much higher near Volcán San Martín Tuxtla on the southwest side. In April 1960, Andres Baxin and I found the fresh tracks of a medium sized tapir at 1600 meters elevation in the mud bottom of the cone inside the crater of Volcán San Martín Tuxtla. In July 1962, I noted tracks of a small tapir at 800 meters elevation in the primary forest north of Volcán San Martín Tuxtla. In late August 1962, Andres Baxin saw tapir tracks measuring about six inches in diameter at about 1100 meters on the south slope of this Volcán. In the same month I found tracks of at least two animals, one also measuring six inches across, near the Río Carizal about two kilometers inland from the Gulf of Mexico.

Pedro Mateo of Piedra Labrada reported some tapirs now, but more in the past. In the coffee plantations on the south slopes of Volcán Santa Marta, Jeotulio Gutierrez said they were "abundant" with trails, tracks, and droppings commonly seen. The older people in this area said that the tapir "is made out of all kinds of meat" and will not eat them for fear of eating horse or burro meat. It appears that most of the remaining tapirs in the Sierra inhabit wilderness areas on Volcán San Martín Tuxtla, between there and the Gulf to the north and northeast, and the peaks and Gulf slopes, river gorges and valleys of the Volcán Santa Marta massif (Figure 9). Although a number of tapirs are killed by the local people on sporadic hunts or on chance meeting, some are taken by organized parties from San Andrés Tuxtla and other larger towns, as well as by hunters from outside the region.

The status of the *jabali* or Collared Peccary *(Pecari tajacu)* in the Sierra is not clear. I saw none during my field work, but reports indicated that they are fairly common locally. Since a large part of the region is covered by heavy humid forest, not ordinarily optimum habitat for this species, it is likely that the Collared Peccary is more numerous in the drier areas inland and in the forest edge and thickets in cleared sections at lower elevations on the Gulf slopes. They are hunted irregularly. I was informed that they are seen in numbers in the La Victoria section as well as in the lowland forests and thickets in the vicinity of Bahía Sontecomapan. They were also reported from the Ocotal area and were considered more numerous near the Gulf of Mexico near Piedra Labrada.

The White-lipped Peccary or *marina (Tayassu pecari)*, as it is known locally, was formerly an abundant animal in the Sierra de Tuxtla. Jeotulio Gutierrez' father, who is less than fifty years old, remembered seeing these pigs filing down a trail in a group a hundred yards long in the Volcán Santa Marta area. Argudin recalled seeing as many as 300 animals about 20 years ago near the Río Yougualtajapan. About 15 years ago Andres Baxin observed a large band of marinas around Volcán San Martín Tuxtla. Dalquest (1949:411) stated that in January 1948, he heard vague reports of White-lipped Peccary in the Sierra, but no person questioned had actually seen one. I questioned people closely in all sections about this species. Most distinguished it readily from *Pecari tajacu*. In all instances except two I was told that there were no marinas left. One exception was a man in Colonia Adalberto Tejeda above Coyame who vaguely indicated there were some left near the Rio Yougualtajapan. The other was Jeotulio Gutierrez who said they had been seen recently above San Fernando, west of Ocotal Chico, but they were rare.

The only first-hand record is my observation of two individuals at about 600 meters elevation in the humid forest about 4.5 kilometers southwest of Sontecomapan on April 17, 1960 (Figure 9). These two passed about six meters away and returned as I remained motionless, watching them forage in the leaf litter on the forest floor. Several informants said the peccaries had left the region and moved southward beyond Acayucan, giving the impression that there was some sort of migration. It is clear from the evidence I obtained that few White-lipped Peccaries remain anywhere in the region, and that heavy hunting was probably a major cause of their rapid decline, since extensive areas of suitable habitat still exist.

I observed tracks of the Red Brocket *(Mazama americana)* in many localities up to the higher elevations through the humid forests of the region. Most people considered this small deer common. Their skins are evident in almost every village and are often used as the seats and backs of wooden chairs. These deer are chiefly nocturnal, usually behave secretively and are seldom observed. In 1962, above Dos Amates, Gary Ross saw a Brocket that neither appeared startled nor ran away immediately. Although we found fresh tracks on several night hunts, we did not observe any deer. Pedro Mateo and Jeotulio Gutierrez said the *temazate* is heavily hunted and is decreasing in the Piedra Labrada and Ocotal areas. This is probably true in other sections, such as the south slopes of Volcán San Martín Tuxtla, where they are frequently hunted.

Origin and affinities. The mammals of the Sierra de Tuxtla reflect primarily the transitional character of Central America and southern Mexico between the tropical and north temperate American mammalian faunas (Darlington, 1957:334). Most families in the region are primarily of North American or Old World origin, but several, Didelphidae (opossums), Myrmecophagidae (anteaters), Dasypodidae (armadillos) and Mustelidae (mustelids), had major secondary radiation in South America. Apparently only the Erethizontidae (porcupines) and the Dasyproctidae (agoutis, pacas) have had their principal radiation very early from South America with no known early ancestors in North America. The origin of the American Cebidae is not certain. They may have dispersed through North America despite no fossil evidence there, but they possibly may be products of parallel evolution in South America. The earliest movements of the bats, a major component of the Sierra's fauna, are unknown as most lack a fossil record. The Emballonuridae (sac-winged bats), Desmodontidae (vampire bats) and Natalidae (funnel-eared bats) are tropical in distribution; the Phyllostomidae (American leaf-nosed bats) are both tropical and warm temperate, and the Vespertilionidae (vespertilionids) cosmopolitan.

Chapter 4

HUMAN ELEMENT

4.1   Settlement and Population

Preconquest period. There is evidence that parts of the Sierra de Tuxtla were settled at a very early date. Figurines considered to be from the Preclassic Culture period (ca. 1500 to 500 B.C.; Sears, 1952:245) have been found in archeological sites in and at the base of the range. Melgarejo Vivanco (1960:19) indicated that coastal sections of Veracruz have been inhabited since at least 1500 B.C. Spinden mentioned (1943:153) that the cradle of the Mayan culture may have been in this coastal belt where arid and humid conditions exist side by side and where such figurines are found together with those of the Mayas. I show in Figure 10 the archeological sites which have been found in the Sierra and their areal relationship with those in the surrounding lowlands. The sites discovered in the Sierra are mostly the ceremonial type with temple and burial mounds, courtyards, stelae and idols. Potsherds, utensils, stone implements and worked obsidian have been found in various localities. The large number of archeological sites at lower elevations in the Sierra is an indication that there was a fairly dense preconquest population, especially during the centuries immediately preceding the arrival of the Spaniards. Figure 10 reveals that most of the sites are around Lago Catemaco and in the valleys to the westward near San Andrés Tuxtla and Santiago Tuxtla. There are several sites in the coastal areas, indicating the presence of isolated early

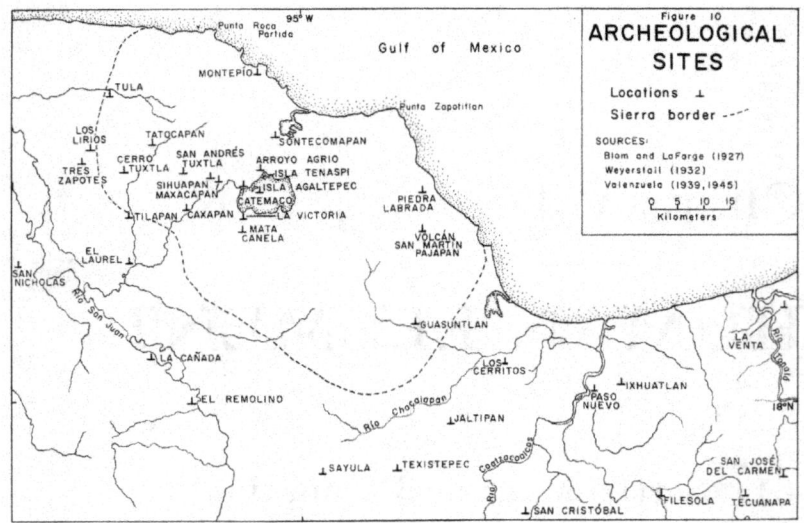

settlements at favorable locations. Preconquest settlements, therefore, seem to have been mostly in the less humid areas on the southern side of the range in locations where fertile soils and good sources of water occur. Apparently most settlement did not extend much above sea level on the coast or above 600 meters on the inland side of the Sierra.

Very few of the archeological locations in the Sierra de Tuxtla have been excavated, so most of the artifacts discovered have been found on or near the ground surface. The Tuxtla statuette, reported to have been plowed up near San Andrés Tuxtla (Holmes, 1947:691), possesses glyph characters which have been interpreted to be Mayan and to contain one of the oldest dates (98 B.C., Spinden, 924; 162 A.D., Morley, 1956:48) in the hemisphere. There is controversy, however, in regard to the statue's origin and whether it was carved later than the apparently early date appearing on it (Blom and La Farge, 1926:42). Valenzuela (1945:90) reported on the superficial excavation of several mounds on the Río Tuxtla at a site known as "La Mechuda" southwest of Santiago Tuxtla. Here he found some small heads which he considers archaic in

100

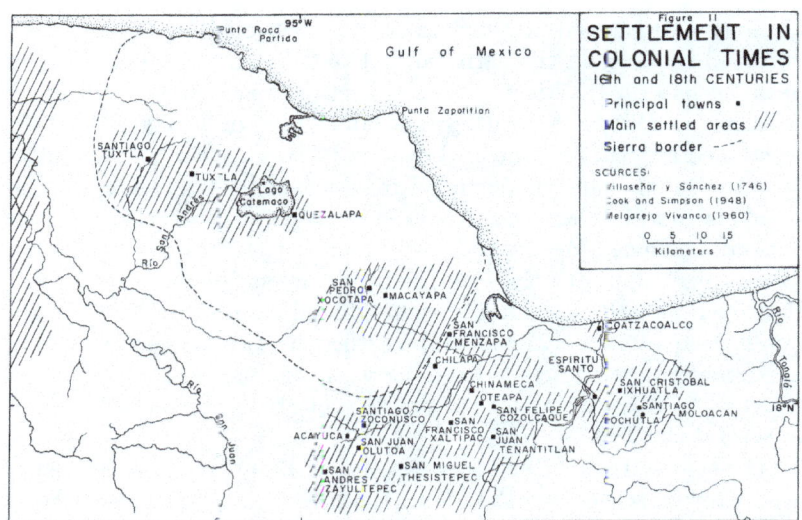

Figure II

SETTLEMENT IN
COLONIAL TIMES
16th and 18th CENTURIES

Principal towns •
Main settled areas ///
Sierra border ---

SOURCES:
Villaseñor y Sánchez (1746)
Cook and Simpson (1948)
Melgarejo Vivanco (1960)

age.

Valenzuela (1939, 1945) investigated sites in the barrio de Campeche (suburb of San Andrés Tuxtla), Matacapan, Isla Agaltepec, near Tatocapan, Pollinapan, Cerro Tuxtla, Mata Canela, La Victoria, Belem Chico (suburb of San Andrés Tuxtla), and Finca Ciruelo (3 km. east of San Andrés Tuxtla). Near Tatocapan he found artifacts with similarities to the Huasteca culture in the Panuco region, and at Pollinapan, also near Tatocapan, designs similar to those of the second epic at Monte Alban, Oaxaca. He concluded, however, that in the region "the most characteristic and abundant are elements of the great Mayan culture." Valenzuela estimated "some 60 earth mounds" near Tatocapan, an indication that this locality was probably densely populated.

In his investigations of the numerous sites outside the Sierra in the vicinity of Hueyapan and the Papaloapan and Tesechoacan Rivers, Weyerstall (1932:30) made the observation that in this region several people of different cultural stages met, and remains of them often occur in the same locality. This appears to be the case in the Sierra as well.

101

Some antiquities from the Sierra and the surrounding lowlands were obtained by Seler in the early part of this century and described by Seler-Sachs (1922). Most of this collection was obtained by gift or purchase and little excavating was done. The only Sierra localities mentioned are San Juan de los Reyes and Mata Canela. As the author indicated, little can be concluded from this collection because of its small size and the lack of systematic excavation. Nevertheless, Seler-Sachs did comment on the styles of the various objects as being similar to cultures from both the south and northwest (Maya, Totonac and Huasteca) as well as to that of the highland Aztecs and Toltecs. She raised the question (1922:547-548) whether the Tuxtla region was on an important travel and trade route between the northern and southern regions, thus accounting for the presence of objects apparently related to several cultures.

During the Tulane University expedition, Blom and La Farge visited a number of archeological sites in the Sierra and mentioned others. They observed some small mounds at Siguapan (Sihuapan), a group of very large ones at Natacapan (Matacapan), several near Catemaco, Mata Canela, Piedra Labrada and on Isla Agaltepec in Lago Catemaco (Figure 10). They heard of others at Tula, Montepío and Sontecomapan. On Agaltepec there are also courtyards, terraces and a truncated pyramid, the whole island appearing to have been modified by man at one time or another. Blom and La Farge (1926:23) said that the stone boxes, altar and serpent's head at Mata Canela look very Aztec but that they believed them to be connected more closely with the Totonac culture.

Although Kerber (1882:488) did not say where the archeological site he referred to is located, except to mention that it was "seven hours from San Andres Tuxtla," it appears from the crude map to be in the vicinity of the present town of Montepío on the Gulf coast. Here, according to Kerber's map, there are at least 19 mounds or structures of various sizes, from two of which came a collection of artifacts including tools for maize grinding and some idols. According to Finck, the collector who visited the locality, the site has characteristics intermediate between Toltec and Aztec cultures, although it is said later the idols are thought to be of Aztec origin. Although there is no map scale, the Montepío site appears to be a fairly large one containing, if Kerber is correct in regard to interpreting dimensions, structures up to 12 meters high and

120 meters in length. This sheltered and well watered location (Plate II) is an excellent one for settlement and no doubt has been inhabited for long periods.

Spinden (1928:651) has a map of archeological ruins in southern Mexico showing sites contemporary with the Mayan Postclassic settlements, the latter apparently reaching a peak about 1200 A.D. In addition to many sites in the lowlands of the Coatzacoalcos and Papaloapan river basins, this map shows at least a dozen localities in the Sierra. Among these are several around and west of Lago Catemaco, and others at San Andrés Tuxtla, Santiago Tuxtla, Montepío, Sontecomapan and Piedra Labrada. Judging from the size of the sites in the Sierra and the cultural diversity of artifacts discovered so far, some of these localities likely have been inhabited more or less continually for over 2000 years. The lowland Olmec sites of Tres Zapotes and La Venta are located near the Sierra (Figure 10). Drucker, et. al. (1959:264-265) listed extreme C-14 dates from La Venta between 1454 B.C. and 126 A.D. (average 814 B.C. as date of Phase I construction). They assume from present evidence that La Venta correlated with Middle Tres Zapotes, and that Lower Tres Zapotes, the earliest Olmec horizon known, ran its course prior to the 9th century B.C. The Upper Tres Zapotes horizon is considered to have ended about 1000 A.D. (Drucker, 1943:122). In view of this chronology, it is possible that some of the Sierra's sites were contemporary with these lowland ones.

Colonial period. The Spaniards entered the Sierra de Tuxtla only a few years after they arrived in Mexico. In 1518, the year before Cortés reached Veracruz, Juan de Grijalva sailed past the mouth of the Río Coatzacoalcos, saw the Sierra, and named the range San Martín after a soldier of that name who accompanied the expedition (Díaz del Castillo, 908:[1]50). The Spaniards first passed through the region while enroute to the Coatzacoalcos area where, in 1522, they established a permanent settlement, Villa de Espiritu Santo (Melgarejo Vivanco, 1960:60).

Although no specific records are available, the Indian population of parts of the Sierra at the time of the conquest was apparently fairly high. This is indicated from Borah and Cook's (1963:81-82) preconquest population figure (based on tributary families in 1568) for the area from Alvarado to the Laguna de Términos, from which it is possible to estimate the population of the Papaloapan and Coatzacoalcos basins at

103

that time as being 575,890. Even though this figure may be excessive, there were certainly many towns in the lowlands surrounding the Sierra, and those within the range occupied much the same areas where most of the archeological sites are located. In Figure 11 I show the major towns and the approximate extent of the main populated sections of the Sierra and adjacent lowlands in the 16th and 18th centuries. The populated area on the western side of the map is that of the prominent lowland colonial towns of Tlacotalpan, Cosamaloapan and Tesechoacan, all of which are situated just outside the map border. Borah and Cook (1963) included a map of the empire of the "Culhua Mexica" (Aztec) which shows Lago Catemaco and the Volcán San Martín Tuxtla massif as the Tochtepec (Tuxtepec) tribute province with Mixtlan, in the vicinity of the present San Andrés Tuxtla, as the only town in the area. Medel y Alvarado (1963:[1]29-30) stated that due to the eruption of Titépetl (Volcán San Martín Tuxtla) in the early 16th century, the residents of a town called Ixtlan (=Mixtlan?) were forced to leave, and possibly founded the town which is the present San Andrés Tuxtla.

Early accounts of the Sierra de Tuxtla during the colonial period are meager. It is estimated that there was a large and dense population in the Gulf lowlands which decreased by one-half between 1519 and 1532. Slaves were reported to be especially numerous in the Gulf coast, perhaps comprising 20 per cent of the population (Borah and Cook, 1963:68-69). Spanish penetration into Gulf coastal areas was uneven. At first there was little effect on the Indian population, then population decreased at an accelerating rate to the end of the 16th century when it began to slacken (Borah and Cook, 1963:89). Foster (1942:7) indicated that the isolated Popoluca Indian populations in the southeastern part of the Sierra were not disrupted to any extent by the Spaniards and were grouped in adjustment to land potential. Some of the conquerors, including Cortés, held land grants in the Sierra (Blom and La Farge, 1926:49; Medel y Alvarado, 1963:[1]32). Cortés established a sugar cane *ingenio* near the coast west of the Río Caña (Medal y Alvarado, 1963[1]40). These were the beginnings of the large landholdings in the Sierra which persisted until broken up in the early 20th century.

In regard to the population decrease noted above, Cook and Simpson (1948:2-3) remarked on the lacunae of towns in the provinces of Alvarado and Coatzacoalcos. There were Indian towns in the early 16th century and later which cannot be identified on modern maps. Most of these vanished communities were in the hot country and the authors suggest that population decrease occurred through exploitation of the Indians on cacao plantations, by slave catching and "senseless devastation," and by Old World diseases. Apparently epidemics in the lowlands became widespread and destructive in the years 1540-1550, 1570-1580 and 1590-1600. Melgarejo Vivanco (1960:126) mentioned a smallpox epidemic in the Tuxtla region in 1580 which caused numerous deaths among the Indians. Cook and Simpson (1948:46-48) traced the population of central Mexico (including the Sierra de Tuxtla) from about 11 million in 1519 to 2.5 million in 1600. Borah and Cook (1953) later changed the 1519 population estimate to 25 million. The decline appears to have continued to an estimated low of 1.5 million in 1650 and by 1793 it had increased to only 3.75 million. Borah and Cock's extremely low estimate of 25,979 as the population of the Alvarado (Papaloapan) and Coatzacoalcos basins in 1568, gives an indication that the decline was drastic. That the Sierra's population also decreased markedly in this century is probable.

In the colonial period there are population figures available for only a few towns in the Sierra and around its base. There is a report that the villa of Santiago Tuxtla contained 22 towns and 22,000 "vassals" of Cortés in 1525 (Medel y Alvarado, 1963:[1]40). Cook and Simpson (1948:120-130) estimated the population of Tuxtla (=Santiago Tuxtla?) as 4000 in 1565. They give the population of Chinameca as 880, Menzapa 2400, Chilapa 920, and Coatzacoalcos 7600 in the same year (Figure 11). Villaseñor y Sánchez (1746:366) mentioned the number of families in three towns in the southeastern part of the Sierra just before the middle of the 18th century. Using an average of four and one half persons per family a population estimate would be Soteapan (San Pedro Xocotapa) 1611, Mecayapan (Macayapa) 482, and Minzapan (San Francisco Menzapa) 284. Acayucan at this time had a population of 1695 (converted estimate), and other towns had populations as follows: Soconusco (Santiago Zoconusco) 1327, Oteapa 310, Jaltipan (San Francisco Xaltipac) 436 and Sayula (San Andres Zayultepec) 630. Paso y Troncoso (1905:5) mentioned that the town of Tuxtla in 1580 had six *estancias* (Conchihca, Sant Andres Zacualco, Matlacapa, Caxiapa, Chuniapa and Catemaco) but he gave no population figures.

The gradual population increase which took place in the Sierra during the 18th century continued into the modern period. It was in the latter part of the colonial period that towns in the Sierra began to attain wider prominence as commercial and political centers. Although Acayucan, lying outside the Sierra proper, and San Andrés Tuxtla were mentioned as important large towns early in the period, they, and especially Santiago Tuxtla, became more prosperous in the late 18th century (Melgarejo Vivanco, 1960:106). San Andrés, however, in the modern period gradually eclipsed all the towns in the Sierra in size and importance as the principal urban center.

Modern period. During the turbulent times of the independence movement in the early 19th century, several laws were enacted in Veracruz which were intended to further the proper disposition and use of land in the state. Important among these were the *Ley Agraria del Estado* and the *Ley de Colonización* of 1826 (Melgarejo Vivanco, 1960:159-160). Medel y Alvarado (1963:[1]311) mentioned that on May 25, 1886, by Decree No. 29 of the Veracruz legislature *terrenos baldíos* (idle lands) were distributed to *campesinos* in Canton Los Tuxtlas. Although some action was taken during this century in regard to setting aside communal land for the poor, for retired military personnel and for colonization, the latter particularly in the Jalapa and Coatzacoalcos areas, no strong efforts were made to break up the large landholdings until after the revolution.

The population of the Sierra during the first half of the 19th century was increasing, as it was in other parts of Mexico, after the low point reached in the middle of the 17th century. By 1850, however, it probably was not more than about 15,000. With the continual population increase there was an expansion of settlement and clearing of forest for agricultural purposes, particularly in the late 19th and early 20th centuries when new plantations were established on the southern side of the Sierra. Population in the cantons of Acayucan and Los Tuxtlas in the last two decades of the 19th century showed a marked increase due to foreign immigration, the rising natural increment and movement from other parts of Mexico. By 1910 the Sierra's population was about 65,000. Table V shows population figures for the two cantons through this period. A small part of the Canton Los Tuxtlas and over half of Acayucan lie outside the Sierra. Therefore, the totals for Acayucan are about 50 per cent greater than the actual population at that time in the part of the canton covering the range.

106

TABLE V			
POPULATIONS OF CANTONS LOS TUXTLAS AND ACAYUCAN AND THE SIERRA DE TUXTLA 1868–1910*			
	Los Tuxtlas	Acayucan	Sierra de Tuxtla (estimated)
1868	21,345	16,559	24,000
1882	28,099	20,563	33,000
1900	43,824	38,164	57,000
1910	48,823	44,451	65,000
* Datos Preliminares del Censo de Población de 1960, Gob. del Edo. de Ver.			

By the end of the revolution and the creation in 1914 of the *Comisión Agraria del Estado* (Melgarejo Vivanco, 1960:208), the division of the large estates was implemented, and the enactment of national laws concerning *ejidos* and small private holdings furthered the settlement of the Sierra de Tuxtla. The new *Ley Federal de Colonización* of 1926 marked the beginning of a period of settlement expansion in Mexico which, however, did not exert a marked influence in the Sierra until more than a decade later.

In 1917 the state was divided into *municipios* and the old system of cantons, created in 1824, was discontinued. Figure 12 shows the approximate boundaries of the nine municipios included wholly or partially within the Sierra. In Table VI are their populations in the three decades from 1930 to 1960. The increase in the population of several municipios in the two decades before 1960 is a reflection of several factors. The completion of the paved road to San Andrés Tuxtla from Veracruz and then to Acayucan during this period brought an increase in commercial and agricultural activity in these municipios with a resultant population upturn. San Andrés Tuxtla had a population of about 10,000 in 1923 (Friedlaender, 1923:165), and this increased to about 22,000 by 1960. Santiago Tuxtla (Plate XXII) and Catemaco, the other two large towns, now each have well over one-third the population of San Andrés. Another factor influencing population is the availability of unoccupied land for settlement. The actual decrease in population of Municipio Santiago Tuxtla from 1950 to 1960, after a large increase in the previous decade, may be partially the result of most all the

107

available land being occupied. The great increment from 1950 to 1960 in Municipio Catemaco's population is partly due to the availability of new land as well as to the town and lake's easier accessibility by new roads constructed during that period. A contributing factor here is the unlimited fresh water supply of Lago Catemaco. The municipios of Soteapan and Mecayapan are still not easily accessible. They also contain large areas where soils are not well suited for agriculture. Consequently, they have not shown as great fluctuations in population.

Other factors have caused the region's total population to change markedly during the last thirty years. Between 1910 and 1930 the Sierra's total population showed no appreciable alteration. Actually the population census in 1921 recorded a decrease in Mexico's population (Whetten, 1948:23), although this trend may have been partially due to less coverage than in other censuses. Since about 1930, however, when the Sierra de Tuxtla's population was about 60,000, modern improvements in health and sanitation methods have reduced the country's death rate. Hospitals (a modern one in San Andrés Tuxtla in the last decade), introduction of new drugs, and more resident doctors in the larger towns of the Sierra have furthered this trend. From 1930 to 1942 the excess of births over deaths in Mexico reached points where it was two to three times that of the United States and more than in most Latin American countries (Whetten, 1948:26). Health and sanitation improvements have still not reached effectively into the smaller and isolated settlements of the Sierra as they have into and near the three largest towns. The malaria eradication program, however, has contributed significantly to health in the area. With support from the United Nations this program has been extended to all parts of the Sierra and has been especially effective in some coastal areas where malaria incidence is high.

Table VII reveals a much higher increase in population in the Sierra de Tuxtla region from 1930 to 1950 than in Veracruz state or Mexico in that period. The region's increase from 1950 to 1960 was only slightly more than that in Mexico and about the same as that in Veracruz. Although there were many foreign immigrants to Mexico from 1928 to 1934, a significant decline occurring afterwards (Whetten, 1948:26), the Sierra did not receive enough to cause such a pronounced increase in population in addition to the high natural increment. The population addition from 1930 to 1960 was the result of an influx from other

108

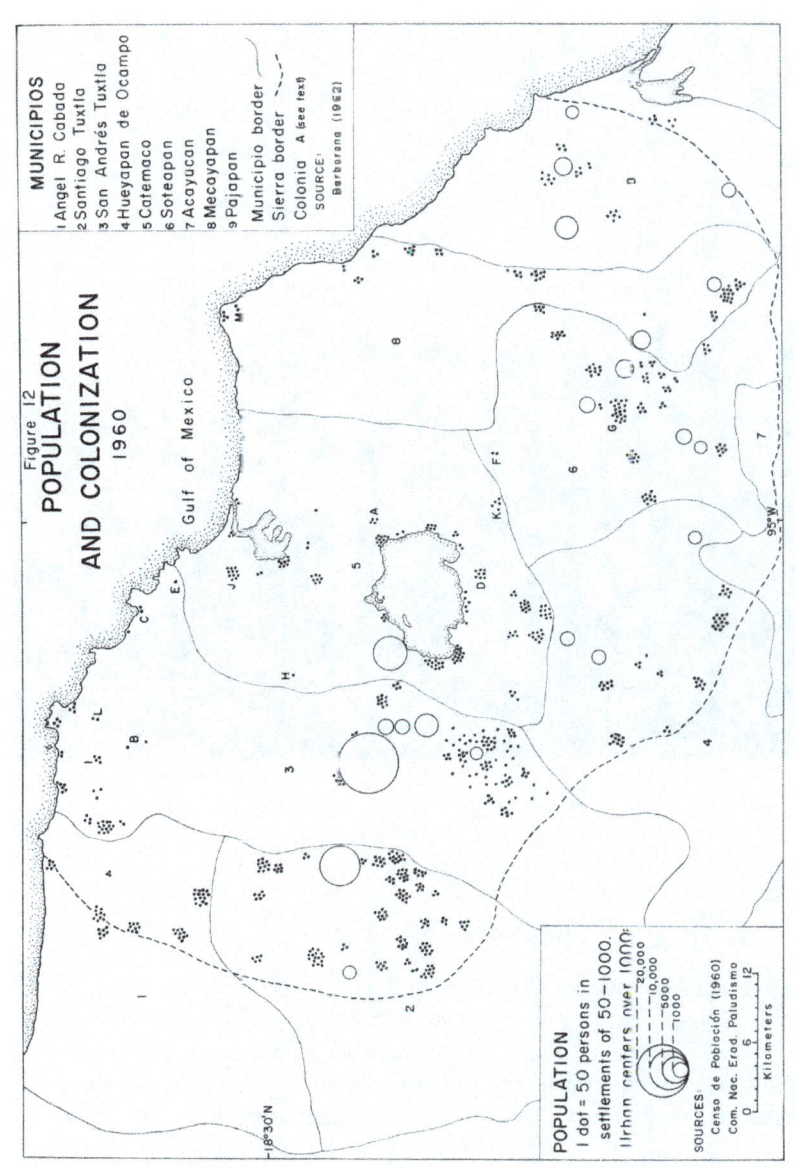

Figure 12
POPULATION
AND COLONIZATION
1960

Gulf of Mexico

MUNICIPIOS
1 Angel R. Cabada
2 Santiago Tuxtla
3 San Andrés Tuxtla
4 Hueyapan de Ocampo
5 Catemaco
6 Soteapan
7 Acayucan
8 Mecayapan
9 Pajapan

——— Municipio border
- - - - Sierra border
Colonia A (see text)
SOURCE:
Barbarena (1962)

POPULATION
1 dot = 50 persons in
settlements of 50—1000.
Urban centers over 1000:
20,000
10,000
5000
1000

SOURCES:
Censo de Población (1960)
Com. Nac. Erad. Paludismo

0   1   2
Kilometers

18°30'N

95°W

109

Plate XXII. Santiago Tuxtla

9 km. west-northwest San Andrés Tuxtla, 200 m. a.s.l. One of the largest urban concentrations, this old town lies in a valley at the base of Cerro Tuxtla. It is an agricultural and commercial center for a long-settled part of the range, but has not shown the growth and activity of nearby San Andrés. Cerro Blanco is visible at upper right.

TABLE VI							
POPULATION OF SIERRA DE TUXTLA REGION MUNICIPIOS 1930–1960*							
Municipio	Approx-imate Area (km2)	Popula-tion	Popula-tion	Popula-tion	Popula-tion	Per cent In-crease	Per cent In-crease
		1930	1940	1950	1960	1940-1950	1950-1960
San Andrés Tuxtla	971	22,639	27,372	44,950	58,314	64.2	31.9
Santiago Tuxtla	507	19,590	16,064	23,871	24,644	48.6	3.2
Acayucan	ca. 525	11,811	9,317	13,188	23,176	41.5	75.7
Hueyapan de Ocampo	616	4,642	9,989	12,652	18,142	26.6	43.3
Catemaco	557	5,107	7,784	8,713	15,962	11.9	83.2
Angel R. Cabada	693	-	8,484	11,182	14,840	31.8	32.7
Soteapan	525	3,496	4,966	6,266	8,520	26.1	35.9
Mecayapan	398	3,807	4,579	6,065	7,357	32.4	21.3
Pajapan	347	3,741	3,554	5,540	5,920	55.8	7.4
Totals	5139	64,833	92,109	132,427	176,875	Av. 43.7	Av. 33.5
* Datos Preliminares del Censo de Población de 1960, Gob. del Edo. de Ver., Comisión Nacional de Eradicación de Paludismo, and Barberena (1962).							

111

parts of the country particularly into the municipios of San Andrés Tuxtla, Santiago Tuxtla and Catemaco. The population increases in municipios Acayucan and Hueyapan de Ocampo, from the same cause, include a large percentage outside the Sierra. The percentage increase in the region's total population from 1950 to 1960 has been lessened to only slightly more than the natural increment of the nation chiefly by the marked decrease in only two of the nine municipios, six of the remaining seven still equaling or exceeding the national percentage.

TABLE VII			
SIERRA DE TUXTLA REGION POPULATION INCREASE COMPARED WITH VERACRUZ STATE AND MEXICO 1930–1960*			
Population Increase			
(per cent)			
Unit	1930-1940	1940-1950	1950-1960
Sierra de Tuxtla Region	42.1	43.7	33.5
State of Veracruz	17.6	25.9	33.7
Mexico	18.7	30.7	31.9
* Calculated from data in Table VI, and *Mexico* in "The Statesman's Year Book" (1933, 1943, 1953, 1962).			

Thus during the past three decades this significant population increase has involved an expansion of urban residence, especially in and around the three largest towns, and an extension of settlement into previously uninhabited or sparsely inhabited places in the Sierra. These latter population thrusts have been in the Catemaco basin, toward Bahía Sontecomapan and in the northwest part of the range eastward from Laguna Majagual. A good example of internal population shifting is the movement of people from the Ocotal area on the southern side of the Volcán Santa Marta massif to settlements with better agricultural land on the Gulf slopes. Ocotal Grande had recently lost population in this way. As early as 1923 Blom and La Farge (1926:42) mentioned a small settlement near the Piedra Labrada ruins which they referred to as a Popoluca outpost, the inhabitants having come from Ocotal Grande. Similar minor internal movements are taking place in various parts of the Sierra, particularly where new roads are being constructed.

In analyzing the Sierra de Tuxtla with regard to total population and density it must be remembered that portions (to a maximum of about 30 per cent) of five municipios lie outside the range. Adjusting the areas and 1960 populations of these municipios accordingly, the total area of the Sierra de Tuxtla is about 4500 square kilometers and the total population approximately 145,000. The population density is about 30 persons per square kilometer (ca. 78 persons per square mile). Although this is a high density compared to most areas of Latin America, the actual distribution of people in the Sierra is uneven. Figure 12 reveals that the majority of the population is concentrated in the southwestern part of the range, chiefly westward from Lago Catemaco to the vicinity of Santiago Tuxtla. Here live over one-third of the Sierra's people. There are several fairly large urban centers on the southern slopes of the Volcán Santa Marta massif. These heavily populated sections are the same areas where population has been concentrated during and since preconquest times. The remainder of the population is irregularly distributed mostly on the southern slopes of the Sierra, in the Lago Catemaco-Bahía Sontecomapan section, and on or near streams or springs where a continual supply of water is available. A comparison between Figures 3 and 12 shows that the areas with the sparsest population, or which are uninhabited, include principally the higher elevations on the large volcanoes, their Gulf-facing slopes and some coastal areas.

In many places, especially in recently settled areas *(colonias)* and more remote localities, there are scattered single houses (Plate XXIII) or groups of several houses *(ranchos)*. About 55 per cent of the Sierra's people live in settlements ranging in size from two or three to two hundred or more houses, with inhabitants numbering from a few to 1000 (Plate XXIV). About 21 villages and towns in the Sierra contain over 1000 persons, the largest being Pajapan (ca. 3800), Tatahuicapan (ca. 3200), Comoapana (ca. 3000), Soteapan (ca. 2300) and Mecayapan (ca. 2200).

4.2   Land Systems and Land Use

Land in the Sierra is divided among three general categories, ejido, private and national. Up to 1944 total land distributed under the agrarian program in Veracruz was 984,331 hectares (Whetten, 1948:140), or about 14 per cent of the state's area. This is one of the highest amounts distributed in the states in southern Mexico exclusive of the Yucatan peninsula. In 1940 there were far more ejidos (1383) in Veracruz than in any other state. A large proportion of the Sierra's land area is under this system. Land maps that I examined showed a kaleidoscopic pattern of ejidos of varying sizes and shapes intermixed with almost equally variable plots of private holdings. Whetten (1948:241) indicated that many ejido boundaries were not settled, and precise surveys are probably lacking in more remote and relatively inaccessible sections of the Sierra. Many smaller private land parcels are marked *pequeñas propiedades* on the land maps, indicating the residual holdings of expropriated haciendas. According to Tannenbaum (1929:507), privately owned rural lands in Veracruz (including many large haciendas) occupied 7,137,000 hectares in 1923, all but 52,600 hectares of the state's total area. Since 1930, however, much land has been expropriated, and there has been a great impetus in the Sierra in the growth of ejidos and especially in small private holdings and colonias. The newer ejidos have principally been created from national lands *(tierras nacionales)*, and occasionally the small landed estates, but initially they were established from expropriated haciendas. The principal areas of national land remaining in the Sierra are in the higher elevations, particularly on the four largest volcanoes.

In Veracruz state about 223 colonias have been established, some from unused national lands *(terrenos baldíos)*, others from private holdings (Melgarejo Vivanco, 1960:212). From 1916 to 1943, however, only three colonias were created in the state (Whetten, 1948:590). Thus, most colonias in Veracruz are less than thirty years old and many in the Sierra are only about half that age. The Department of Agriculture and Livestock determines whether land is suitable for colonization. Colonias contain only private land, are administered by the government until 50 per cent of the purchase price is paid, when control is given to the members (Tannenbaum, 1929:285-286). Both colonias and ejido lands are subject to the regulations of the Forest Law. The colonias in the Sierra vary in size, but some are fairly large, covering 5000 hectares or more. Following is a table listing most colonias that have been created in the Sierra and their approximate population at the present time. A letter after each corresponds with a

114

Plate XXIII. Pioneer Settlement

Near Colonia Magdalena, about 700 m. a.s.l. This aerial photograph shows a recent slash-burn clearing of a few hectares in the montane rain forest. Typically, a small number of trees have been left standing on the gently undulating terrain.

Plate XXIV. Mountain Village

4 km. southeast Volcán Santa Marta, 580 m. a.s.l. This view of Ocotal Grande is typical of most small villages in the Sierra de Tuxtla. Here there were about 20 houses and 300 inhabitants, but some people have moved to the north in order to work new land. In the background is the south crater wall of Volcán Santa Marta.

116

letter on Figure 12 showing its location. The establishment of these colonias is part of a federal plan to encourage settlement of unused land in the southeastern part of Mexico and elsewhere.

TABLE VIII		
SIERRA DE TUXTLA COLONIAS*		
Colonia	Map Letter	Population
Adalberto Tejeda	A	185
A. Ruiz Cortines	B	90
Agrícola Montepío (Militar)	C	110
Aguila	D	375
Balzapote (Militar)	E	50
Bastonal	F	75
Benito Juárez	G	1015
Hidalgo	H	-
Huatusco	I	180
La Palma	J	540
Magdalena	K	230
Nuevo o Ruiz Cortines	L	45
Perla del Golfo	M	95
* Comisión Nacional de Eradicación de Paludismo.		

Agriculture, grazing and forest use. In the early part of the preconquest period the people of the Sierra practiced subsistence agriculture as well as hunting, fishing and gathering in a largely forested region. Melgarejo Vivanco (1960:19) indicated that some of the early inhabitants on the coast of Veracruz attained a degree of prosperity, but it is probable that there was a rather low level of subsistence in most areas. The Sierra is favored in possessing large sections of fertile volcanic derived soils, so the people in most of the densely settled places on the southern slopes were able to support themselves more easily than was possible in less suitable Gulf coast locations. Although many people in the Sierra, especially in the isolated coastal settlements bordered by heavy forest, continued migratory tillage, in the densely populated areas a more sedentary agriculture developed later in the period. Maize was the principal food crop grown in both the slash-burn and sedentary types of agriculture. Since forests were more extensive

117

and game more abundant in preconquest times, the people utilized these resources to a greater extent than they do now. The list of tribute items in the Aztec period (Borah and Cook, 1963:Table 1) (Paso y Troncoso, 1905:5) for the Tuxtepec province (which included the Sierra) mentions such products as cotton, latex, liquidambar and cacao. Cotton growing was of considerable importance in the Sierra in preconquest times, was still one of the principal crops in the middle of the 18th century (Covarrubias, 1946:28), and remained important until the latter part of the 19th century (Melgarejo Vivanco, 1960:216).

Despite the population decrease after the conquest, there was an increase in planting, in addition to cotton, such crops as cacao, rice, vanilla beans, peppers and tomatoes. Oranges, Spanish beans and sugar cane were introduced. Maize, however, remained the staple crop of the Indians. Melgarejo Vivanco (1960:82) mentioned that four maize crops annually were produced in Tuxtla and Acayucan; and Villaseñor y Sánchez (1746:366) referred to the same occurrence in Acayucan. During the colonial period the introduced horses, mules and cattle, as well as pigs and chickens, greatly increased as both sedentary and migratory populations depended more on these sources of transportation and food. With the population increase later in the colonial period and the implementation of land distribution during the modern period, agriculture and grazing expanded in the Sierra, especially on the southern side of the range. These are the two major activities in most of the unforested land. Permanent subsistence agriculture is at present dominant in unforested areas. Slash-burn cultivation is concentrated along the edges of unbroken forest, occasionally well within it and also in the colonias. In many areas it is being replaced by permanent cultivation as the population increase precludes land abandonment and forest regeneration. Actual migratory tillage, although common in preconquest times, is decreasing until currently it is engaged in by relatively few people. The expansion of plantation crops during the colonial period occurred mainly around the larger towns, chiefly San Andrés Tuxtla, and at lower elevations on the southern slopes of the range.

During the 19th and early 20th centuries a number of German immigrants established tobacco and sugar cane plantations, mostly near San Andrés, between there and Catemaco and around the lake. As early as 1850, a cigar factory was established in San Andrés Tuxtla

118

(Medel y Alvarado, 1963:[1] 362). The importance of tobacco in the Sierra between 1894 and 1912 is brought out by Medel y Alvarado, (1963: [1]366) who stated that there was a "volume of 15 or 20 thousand bales of raw tobacco of 92 kilos each, annually produced in the tobacco plantations of the Canton Tuxtla." Ninety per cent of the local population of San Andrés were farm laborers cultivating maize, rice, coffee and sugar cane, tobacco absorbing the greater quantity of workers and coming to be the economic base of the Canton. Santiago Tuxtla, lacking tobacco land, contributed manpower, beasts of burden and other products and created a brisk commerce between the towns. Catemaco had tobacco plantations in Olotepec, La Victoria, Matacapan, Mata Canela and other nearby places (Medel y Alvarado, 1963:[1] 356-357). Many tobacco workers were brought in from the states of Puebla and Oaxaca for temporary work, the greater part of them dying of the climate and physical maltreatment (Medel y Alvarado, 1963:[1] 356-357). These tobacco plantations declined, however, when land reforms accelerated after the revolution and the World War in 1914 prevented marketing abroad (Medel y Alvarado, 1963:[1] 365). Blom and La Farge (1926:21) mentioned the existence of several tobacco plantations managed by Germans along the shores of Lago Catemaco, but by that time the crop had greatly decreased in importance. Most of the tobacco is now raised around San Andrés Tuxtla, Sihuapan, Matacapan and Comoapan. It grows rapidly, is planted in August and harvested in November. The better grade is exported, as it has been since the 19th century, and the poorer tobacco kept for local markets.

The largest areas of plantation agriculture are now at lower altitudes mostly outside the Sierra proper west of Tapalapan and south of the Arroyo Hueyapan. Plantations occur locally around San Andrés, Lago Catemaco and Sihuapan, and in some of the colonias and fincas as well as in forested areas higher on the mountains. The distribution of agricultural types is shown in Figure 13. Though grazing is becoming more prominent and is concentrated in some localities, it is usually represented by small herds of livestock scattered throughout the sections of permanent subsistence and slash-burn agriculture.

A majority of the Sierra's farmers depend mainly on a permanent form of cultivation with many also practicing shifting field agriculture (cf. Type 8; Watters, 1960:65). A small minority depend almost entirely on shifting cultivation and relatively few carry on pastoralism.

119

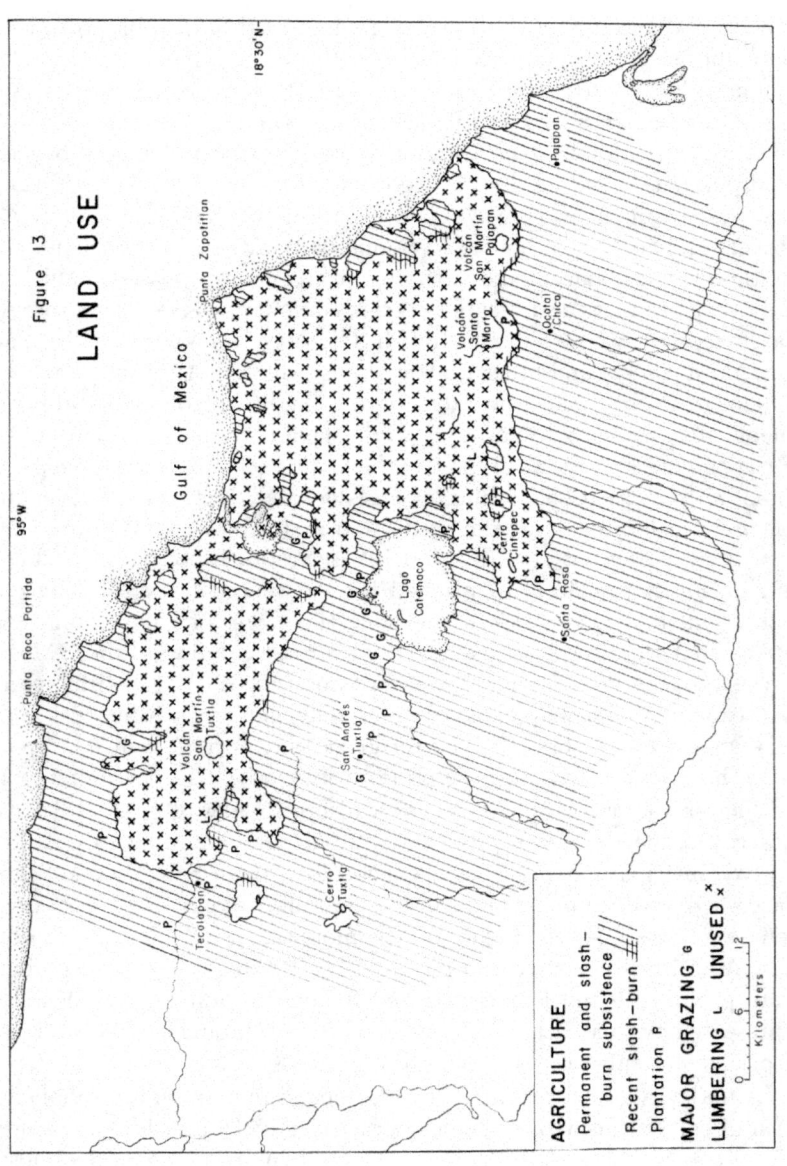

Figure 13

LAND USE

AGRICULTURE
Permanent and slash—
burn subsistence
Recent slash—burn
Plantation P

MAJOR GRAZING G

LUMBERING L   UNUSED x

0   6   12
Kilometers

The "incipient" sub-type of Conklin (Watters, 1960:66), which refers to the beginnings of shifting cultivation by a settler moving into an upland section usually from an overpopulated permanent field area, is evident in the Sierra. There are few places where settlements shift as well as the fields. Where this occurs usually only one or a few houses are involved. In most parts of the Sierra soils are sufficiently fertile so as to preclude migration of whole communities. Frequently temporary houses or lean-tos are built in slash-burn clearings distant from a village. The permanency of most Sierra settlements indicates a higher cultural level than exists where shifting cultivation is a way of life, and marks an environmental relationship nearer to permanent cultivation (Watters, 1960:67).

Maize continues today as the principal subsistence crop in the areas of permanent and slash-burn tillage. Three crops a year are sometimes managed in a few localities, such as at Piedra Labrada (Blom and La Farge, 1926:61), but two is the usual number. The first crop *(temporal)* is planted about May and harvested in November or December, the second planted about December and gathered in May and June. The digging stick and hoe are generally used; oxen and plow are employed in more densely populated areas. Planting is done in rows sometimes with end stakes. The holes vary from about one-third to two-thirds of a meter apart for beans and maize respectively. Usually the hoe is employed to eliminate the fast growing herbaceous plants and weeds. The maize stalks are bent double when nearly ripe for protection against wind and birds. Beans, another major subsistence crop, are planted at various times in the year, occasionally between corn rows. Both these crops yield well in the Sierra, although harvests are generally poorer in the less fertile soils in the pine-oak-savanna area on the southern slopes of the Volcán Santa Marta massif. Tomatoes, melons, pineapples, mangos, squash, chayotes, sweet potatoes, onions, peppers and several other vegetables and fruits are grown in smaller quantities for consumption and market sale. Many of the fields near and particularly to the south of Catemaco, San Andrés Tuxtla and Santiago Tuxtla are intensively cultivated. Though some are allowed to remain fallow, this is usually for not more than two or three years. Cover crops are generally not used and in most sections a rotation of more than two crops (maize and beans) is not apparent.

Slash-burn agriculture in the Sierra follows the well-known seasonal pattern, the forest being cut down by machete and axe during the drier period (February to May) and maize then planted between the burned stumps and downed trees. Occasionally the crop is not sown until the second planting period or sometimes the plot is not burned and planted until the temporal of the following year. If primary forest is available, it is usually chosen in preference to secondary stages. In clearing, the machete is the principal tool used to cut small trees, climbers and lianas, as well as to trim branches from the larger trees. The soft wood of most tropical trees facilitates felling and trimming. Often scattered large trees are left standing. Many small limbs are piled for burning and fires are then set in a number of places within the destroyed forest. Large trunks burn for several days presenting a vivid scene at night. Burning is often not complete, but decomposition of unburned and charred material is rapid. The generally shallow leaf litter and humus quickly dry out when exposed and are consumed by the fire, or undergo rapid disintegration. I examined soil conditions in several areas recently cleared in the humid forest and found little evidence of erosion. The soil was loose and friable, in many locations exhibiting a fine gravelly texture appearing to have a high porosity.

The cropping cycle on newly cleared areas varies in the Sierra. Where the volcanic derived soil is most fertile, up to ten or twelve years of continual use can be obtained, although a decline in the harvest becomes noticeable in the latter part of the period. The time is less in other sections, cycles of three or four to eight years being possible. In the Popoluca country south of Volcán Santa Marta, Foster (1942:17-18) says that the land can be cultivated for five to ten years but is left idle for about the same period, and the abandoned area produces best in the first three years of subsequent use.

The period of abandonment or fallowing is variable in the Sierra. It may range from two or three years to between five and ten, depending on local circumstances. The lack of sufficient land to support an individual or family may force a shortened period of abandonment. In some cases the rapid growth of weeds, other herbaceous plants and shrubs causes abandonment, particularly where sufficient manpower or time for cultivation is not available (Plate XXV). Thus, a combination of circumstances determines periods of land abandonment in the Sierra where families or even villages are concerned. In the wilder parts of the Sierra it appeared that the clearing of new sections in the rain

Plate XXV. Abandoned Rain Forest Clearing

5 km. south Colonia Huatusco, 650 m. a.s.l. A dense, low, herbaceous and woody vegetation is growing up in this recently cleared area. The high rainfall here contributes to rapid plant growth initially consisting of some species not from the original forest.

forest for agriculture, the size of the clearings, and the use or abandonment of those recently cleared, depended more on the inclinations and immediate subsistence needs of the people than upon any systematic plan.

Some newly cleared areas in the range are employed for cattle grazing after harvesting, or are used only for grazing (Plate XXVI). In places burned frequently or where a sufficient number of cattle are grazed, coarse grasses gradually spread and herbaceous and woody plants are prevented from taking hold. In other cleared parts of the forest, especially in locations remote from villages, complete abandonment allows a rapid regrowth of natural vegetation. On the southern slopes, where rainfall is considerably less and soils locally less fertile, this regeneration is much slower and grass areas become more prominent.

The plant succession in slash-burn areas is not only dependent on burning, rainfall and edaphic conditions, but upon proximity to the forest. In clearing at the edge of the Sierra's rain forest or within it, natural seeding from forest plants is accelerated. In such localities the initial growth consists of low herbaceous plants intermixed with woody shrubs quickly forming a dense cover (Plate XXV). Large leaved plants are common at this stage, especially at the forest edge. Species of *Heliconia, Hamelia, Cordia, Conostegia, Polymnia* and *Piper* occur, as well as many others. Some plants at this stage are types found in the rain forest understory, others are quick-growing, light-loving kinds, some being probably brought in by seed-eating birds. Apparently the small palms, common in much of the Sierra's forest, do not regenerate until a later stage. The faster growing shrubs soon eclipse the other vegetation in height so that the clearing presents an uneven appearance. Also common at this stage are nettles *(Urticaceae)* and bamboo. One of the first trees to become established, especially at the forest edge, is *Cecropia*. Occasionally these trees become dominant later in the succession, but most often they are scattered among other species. As growth continues there is a gradual decrease in the density of woody shrubs, herbaceous plants and saplings as forest trees become larger and create more shade. This secondary forest differs radically in structure from the climax. It is more uniform, has no distinct strata and contains a fairly dense understory of shrubs, small trees and bamboo with the larger trees not differing much in size. Climbers, lianas and epiphytes are few. Occasionally there is a giant tree when scattered ones of the

Plate XXVI. Recent Settlement

1 km. south-southeast Colonia Huatusco, 500 m. a.s.l. These two houses are in a section that has been cleared from primary forest and is now used principally for grazing. Parts of the small volcanic cone at right have a cover of secondary vegetation.

original forest have been left standing. The number of tree species in the secondary forest that are not found in the original forest appears to be small. It is higher where patches of forest are surrounded by open or semi-open land and where the forest has been subject to frequent disturbance over a long period of time. Complicating factors in the time required for forest regeneration in the Sierra are the variable terrain, soils and exposure affecting plant growth. The time for a slash-burn area to reach the climax stage of rain forest in this region must, therefore, be variable. I would judge that the time required for a small completely slash-burned area within primary rain forest to reach a climax stage when undisturbed and under optimum rain forest conditions would be at least a century and a half. At the forest edge or at some distance from primary forest, it would take considerably longer. In the latter case the sequence of structural succession would involve a longer weed, herbaceous plant and shrub sequence, more light-loving plants and a later development of the forest stage.

Plantation crops in the region are principally sugar cane, coffee and bananas (Plate XXVII). Also planted for commercial use are tobacco, vanilla and fruit, especially oranges. The cane yields one crop per year and is grown more at lower elevations where there is only a slight gradient. North of San Andrés, however, at Río Grande, there are extensive cane fields at about 600 meters elevation. The cane here is used chiefly to manufacture *aguardiente.* Coffee was first grown in the Sierra in the late 19th century. It is now raised usually in small plots between 500 and 1300 meters elevation in the ejidos, colonias and private holdings, and is shaded by forest trees (Plate XXVIII). There are a few larger coffee fincas, such as Finca La Selva, a plantation of several hundred hectares near Cerro Cintepec. Banana plantations are also usually in small sections except for a few extensive plantings near Chilapan, Cuetzalapan and south of Bahía Sontecomapan. Although bananas yield well, several years ago Panama disease halted the marketing from some plantations, such as the extensive one near Cuetzalapan on the southeast side of Lago Catemaco. Oranges are grown for commercial sale in several locations, the large orchard of the Tegomas in Tapalapan being a good example.

Although cattle have been employed for labor, milk and meat by the people of the region since they were introduced by the Spaniards, the number of animals has been slowly increasing until new herds of

126

Plate XXVII. Cane Cultivation

2 km. west La Victoria, 250 m. a.s.l. The sugar cane in the foreground
is a commercial crop usually grown at low elevations in the range. The
land gradually rises to the unbroken rain forest which covers the small
cerros and Volcán San Martín Tuxtla in the distance.

Plate XXVIII. Coffee Plantation

East slope Volcán Santa Marta, 760 m. a.s.l. The undergrowth of this humid montane rain forest has been cleared out and coffee bushes planted at irregular intervals. Most of the large trees are left to shade the bushes.

twenty to fifty can occasionally be observed near Catemaco and San Andrés Tuxtla. Blom and La Farge (1926:61) reported that the keeping of cattle was rare or unknown among the Indians of the Volcán Santa Marta region. Today there are cattle in this massif, but they are less numerous than in the northwestern part of the Sierra. Primarily cattle are used for milk, but beef cattle are being introduced as more land is utilized for pasture. Unfortunately, the native grasses are generally coarse and nutritionally poor, so in some localities the cattle are not in good condition. Some grass areas are burned annually in the dry season, particularly to the northwest of Lago Catemaco. In places, such as Finca Rociela on the north shore of Lago Catemaco, and in some of the colonias, nutritional grasses are being introduced (e.g., *pangola* from central Veracruz). This, along with importation of cattle types (Zebu or strains of this breed) better suited to the climatic conditions of the region, is resulting in a gradual improvement in the livestock situation.

I observed two localities where commercial lumbering was taking place. In the virgin forest near the Cumbres de Bastonal, about 7.5 kilometers east-southeast of Cuetzalapan, some selective private logging had been going on for several months at an elevation between 750 and 950 meters (Plate XXIX). About five kilometers north-northeast of Tapalapan, trees were being cut in the humid forest in the foothills west of Volcán San Martín Tuxtla at about 400 meters elevation. They were supplying a small sawmill near Tapalapan and the lumber was for local construction. Some trees being cut for lumber in these localities are *chagani, jaboncillo, jiccrillo, laurel, laurellio, palo blanco, palo de bejuco, palo muladoro, palo verde* and *tepozontle* (see Appendix B, Part II). In the regional magazine *Los Tuxtlas*, Argudin (1962:10) condemns the destruction of forests by lumbering on and near Volcán San Martín Tuxtla, and asks the President of Mexico to create a national park in the region to preserve the flora and fauna. Elias González, chief of this forest section, informed me that this cutting commenced in Colonia La Palma, but was now no longer in operation. He said that the only two areas being lumbered at present are those previously mentioned.

Near Tapalapan people told me that some tropical cedar was cut but that there is little present. People in other localities also said that cedar is not common in the Sierra. M. de Escobar (1794) mentions that there is cedar *(cedro)* and mahogany *(caoba)* in the vicinity of

129

Plate XXIX. Lumber Camp

7 km. east-southeast Cuetzalapan, 800 m. a.s.l. This lumbering in the rain forests below the Cumbres de Bastonal is usually selective, and trees are felled in scattered areas of the forest. Oxen are used to transport logs to the sawmill near the camp.

Catemaco "by the other side of the lake," "much cedar" near San Andrés, and also cedar by Tuxtla [=Santiago Tuxtla]. In Jaussoro (1961:5), Gomez Mayorga (1961:10), "Santiago Tuxtla..." (1961:14) and Medel y Alvarado (1963:[1] 356) cedar and mahogany are mentioned as being present in Canton Los Tuxtlas but no specific localities are given. Goldman (1951:269, 283) also reported that both cedar and mahogany occur in the Sierra. Boettiger de Alvarez (1961a:15, 17) said that both types of trees are near Montepío and on the Gulf slopes. My informants indicated that there is now a very small amount of mahogany above Cuetzalapan on the eastern side of the Lago Catemaco basin and also some cedar in the vicinity. They said that most of it is gone and generally only small trees remain. A number of people in various parts of the mountains said that there is mahogany, but some Popoluca Indians indicated that it was uncommon on the Santa Marta massif. I did not identify either species during my field work.

It appears that tropical cedar was at least locally common on the southern slopes of the Sierra in colonial times, and that mahogany occurred less abundantly. The species of these two types occurring in southeastern Mexico are most numerous at low elevations. Beltrán (1961:107) said that in Mexico they are found in the zone below 900 meters altitude. In Guatemala, Standley and Steyermak (1946:448, 458) stated that tropical cedar *(Cedrela mexicana)* is found chiefly at 600 meters elevation or less and mahogany *(Swietenia macrophylla)* mostly below 400 meters, both being most abundant in lowland plains or foothills. It is to be expected, therefore, that they would occur at low altitudes in the Sierra de Tuxtla. These woods have been and still are in demand, so it is probable that they have been especially sought for in the region, particularly after the conquest. In the period between 1894 and 1912, Medel y Alvarado (1963:[1] 358) stated that there were four factories making cedar cigar boxes and two making large crates of soft wood for packing the cigars to be exported abroad. This operation probably removed much of the cedar from around Catemaco and San Andrés Tuxtla. Other than this activity, there is no record of any large scale effort to lumber cedar or mahogany in the 20th century or earlier. The region's inaccessibility, the known low density of mahogany, and the great difficulty in getting the lumber out to markets were probably influencing factors. There may have been some intensive efforts to cut cedar and mahogany in the most accessible places, such as around Lago

131

Catemaco, but it is probable that much of these woods were gradually cut over a long period of time by the Sierra's inhabitants. Most of the forests at lower altitudes on the inland side of the Sierra are not the humid rain forest type in which mahogany grows most abundantly. The less humid conditions and semideciduous character of the forests are more suitable for cedar (cf. Leopold, 1959:32-33). Also, the majority of the forests in these sections have long been destroyed and with them probably went some of the cedar. Although I have a recent report that there is no mahogany in the vicinity of Sontecomapan and Montepío, the extensive forests on the lower Gulf slopes of the range may contain trees of the two species.

Charcoal making, an important occupation in parts of Mexico for centuries, probably has been carried on by some people in the Sierra de Tuxtla for a long time. The extensive forests of the region provide an abundant source of wood for such a purpose. In the late colonial period Escobar (1794) said that the Indians cut wood close to their towns, and that there is much firewood of all kinds, so abundant that "100 towns such as these two [San Andrés Tuxtla and Santiago Tuxtla] are not capable of consuming it" It does not seem likely that charcoal making was practiced by a great number of the Sierra's people in the early centuries, although historical accounts rarely mention it and there is no basis for positive conclusions. Since charcoal's main purposes are a hotter and longer lasting fire, and it is more easily transportable than wood, it is possible that more charcoal was made as the forests were destroyed to greater distances from towns. I do not, however, think that charcoal making reached a degree of magnitude in the Sierra to be a significant factor in the destruction of its forests. The gathering of firewood has been and still is a minor factor in this regard, but the action responsible for almost all the forest removal has been the clearing of land for agriculture.

Few people now are occupied in making charcoal. During my investigations through most parts of the range I encountered charcoal making only once, above Tapalapan (Plate XXX). These men were using various species of rain forest trees with harder woods for its production. I was informed that *chagani* and *palo colorado* were some types used, as well as *encino* and *huisache*, the latter two from the drier forests. The almost universal use of firewood by the rural people and the increase in the use of gas, electricity and oil in the larger urban locations have almost eliminated charcoal production in the Sierra de Tuxtla. In recent years an important factor limiting charcoal production is the forest law (Secretaria de Agricultura y Ganadería, 1961, Art. 38, p. 13; Art. 110, p. 60) which controls the use of fire by *carboneros* and others.

Hunting, fishing and gathering. Before the coming of the Spaniards, hunting and fishing were important occupations of many inhabitants of the Sierra, though these pursuits were primarily supplemental to the basic agricultural activities. Foster (1942:25-26) said that the methods of the Popoluca Indians involved hunts with dogs and with lights at night; sometimes there were communal drives for deer, raccoon, coatimundi and peccary. He also said that guns, bows and machetes were used, as well as nooses and snares for pacas and baited cage traps and slings for birds. Blom and La Farge (1926:56-58) found a revival among the Popolucas in the use of bows and arrows for hunting and fishing, since bandits had come into the region about 1910 and taken their firearms. Apparently by 1925 firearms were being reintroduced and the use of bows and arrows was decreasing. They (1926:44) tell of meeting a man and four boys north of Ocotal Grande cleaning a Great Curassow just secured with a bow and arrow.

That hunting for subsistence has generally declined in importance is evident from the increased dependence of the people on agricultural and domestic animal products and from the reduction in numbers of wild game species. Most hunting is done by people in the small, remote villages near habitats where animals are abundant. In open country there is little hunting because few or no game animals exist; however, some people in the larger towns hunt irregularly for subsistence food and sport. I saw several parties come from outside the Sierra to hunt. The hunting methods Foster mentioned are still used, except the shotgun has replaced the bow and hunting parties are usually composed of one or no more than three or four persons. Near Laguna Tisatal I watched two hunters and a dog pursue an armadillo through heavy forest and then dig it from its burrow with a machete. In some remote areas, such as Colonia Magdalena, the people carry shotguns as a matter of habit in case some animal appears while they are working in their newly cleared land. The high cost of weapons and ammunition is a major factor limiting the number of people able to hunt as well as curtailing frequency of hunts. Most hunters in the Sierra will try to secure any

133

Plate XXX. Charcoal Production

6 km. northeast Tapalapan, 480 m. a.s.l. Four of these men are making charcoal from rain forest trees. Few people are now engaged in this occupation in the Sierra de Tuxtla.

mammal species if they are given the opportunity. The Nine-banded Armadillo *(Dasypus novemcinctus)*, Eastern Cottontail *(Sylvilagus floridanus)*, Gray Squirrel *(Sciurus aureogaster)*, Paca *(Agouti paca)*, Agouti *(Dasyorocta punctata)* and Red Brocket *(Mazama americana)* are particularly sought, but opossums, monkeys, raccoons, coatis, ocelots, tapirs, peccaries and occasionally anteaters and porcupines are also taken.

Species of birds from many families are hunted in the Sierra, but with the exception of the Cracidae, this is done irregularly. The Great Curassow *(Crax rubra)* and the Crested Guan *(Penelope purpurascens)* are two major objects of the hunt. Tinamous, especially the Slaty-breasted Tinamou *(Crypturellus boucardi)*, as well as Spotted Wood-Quail *(Odontophorus guttatus)*, Bobwhite *(Colinus virginianus)* and various duck and dove species are hunted for food. Reptiles and amphibians are not often taken, but iguanas are sometimes secured in the less humid sections, and sea turtles on the Gulf coast.

Fishing for subsistence and marketing is performed along the Gulf coast, in Bahía Sontecomapan and in Lago Catemaco. Inhabitants of the small coastal villages like Toro Prieto frequently supplement their diet with ocean fish. Some of the larger streams and crater lakes occasionally are fished for subsistence food. There appear to be few species of fish generally of medium to small size in Lago Catemaco. Fish inhabiting the lakes and streams are various species of *mojarra* as well as *topote, cuatopote, pepesca, moquille, juile* and *moguille* (Soto, 1961:18). Fishing has been an important occupation of Lago Catemaco since preconquest times. I counted a maximum of 45 fishing boats at once on the lake. Many fish caught here are marketed in Catemaco and San Andrés Tuxtla.

Many of the Sierra's inhabitants gather various types of fruits, nuts and herbs from the forests. Some fruits secured in this way are *tepejilote (Chamaedorea* sp.*)*, *aguacate (Persea* sp.*)*, *guayaba (Psidium* sp.*)*, *chico zapote (Achras* sp.*)* and *nanche (Byrsonima* sp.*)*. Guanabanas *(Annona* sp.*)* are cultivated as well as obtained in the forest. In a long established custom during Easter week (Friedlaender, 1923) people from San Andrés Tuxtla and nearby localities cut large quantities of a fragrant shrub about the peak of Volcán San Martín Tuxtla. I witnessed this event in 1960, when at least 200 people participated in the 25 kilometer round trip to the volcano on foot. The shrub has an

135

odor somewhat like sandalwood, is used for decorative purposes in San Andrés and sold in other parts of Mexico.

Chapter 5

CORRELATION AND SYNTHESIS

5.1 Relationships

Physical environmental influences on the flora and fauna. Prior to human modification of part of the landscape of the Sierra de Tuxtla, its fauna and flora were changing in response to both external and internal influences from the physical environment. Climatic elements, water, soils, volcanism and fire were continually acting separately and in combination to alter the ecological status of plants and animals. A generally mild and humid climate in the region during the Mesozoic and most of the Cenozoic Eras was conducive to vegetation growth and an environment well suited to faunal populations. The series of marine inundations during these eras partially destroyed plant and animal life. Some volcanic disturbances accompanying the Sierra's uplift in the Tertiary were characterized by explosive eruptions and ash falls. Others involved lava flows, some extending as far as 15 kilometers. Destruction of forests and animals was considerable during such times with possibly whole populations eliminated. A slow recolonization by plants and animals took place as volcanism lessened and periods of weathering disintegrated the lava flows, allowing vegetation to develop and provide habitats.

Slightly cooler temperatures occurred at times during the Tertiary, and the general decrease of the Sierra's average temperature in the late Pliocene and in the Pleistocene was apparently small (Dorf, 1959:183, 195). Although in these cooler periods there were some changes in vegetation involving a decrease in tropical and an infiltration of northern forms, the basic character of a humid tropical and subtropical well vegetated mountain region was essentially maintained. The proximity of the Sierra to the Gulf of Mexico has contributed to its mild climate and perpetuation of relatively stable climatic conditions. The temperate elements in the vegetation (e.g., species of *Liquidambar*, *Ilex* and *Viburnum*) of these montane forests indicate a former connection with northern forests that may have been pre-Pleistocene (Martin and Harrell, 1957:478). A slight shifting of faunal elements probably took place with lowering of temperature in the Cenozoic and Pleistocene, this exchange being marked by immigration of forms from the north and an evacuation southward of tropical species. The strong tropical affinity of the region's fauna, however, probably persisted through most of its history. A descent in elevation of subtropical conditions during cooler periods and a final retreat to higher levels above about 700 meters isolated both floral and faunal elements and permitted endemism to develop among some groups of the latter. Since the Sierra lies near the southern border of the Nearctic zoogeographical region, it is open to both northern and southern movements of animals through the filter corridor of Central America. Thus, its faunal fluctuations have been partially independent of environmental regulations, particularly in regard to highly mobile animals of widely ranging families.

After the Pleistocene there were probably no abrupt modifications in the flora and fauna by physical elements other than volcanism and hurricanes. Volcanic disturbances, though less severe and more restricted in area than during the Sierra's formation, continued to destroy animal and plant life possibly including some endemic forms. That such activity had serious effects on the flora is indicated by Mociño Suarez de Figueroa's mention (Medel y Alvarado, 1963:[1] 120) of fires and burned tree trunks caused by Volcán San Martín Tuxtla's 1793 eruption.

The prolonged weathering and the resultant decomposition of fresh volcanic material into soluble mineral components combined with high temperature and abundant precipitation in post Pleistocene times to support a luxuriant growth of plants in many parts of the Sierra de Tuxtla. Changes in this condition are now evident on the south side of the Volcán Santa Marta massif where less rainfall, older and deeply weathered soils with decreased plant nutrients support less luxuriant plant formations (pine-oak) which are low in density and often in height. Plant destruction from lightning caused fires has been likely a minor influence in the Sierra's history. The resistance of the humid forest to burning and the occurrence of electrical storms mostly during the rainy season would have restricted fire from this cause chiefly to the drier southern side of the mountains.

Except for hurricanes, no meteorological phenomena have affected seriously the plants and animals in the Sierra. A few hurricanes have passed near enough to cause forest destruction by winds and loss of animal life from flooding, but the Sierra de Tuxtla is outside the major path of these tropical storms. Frequently, strong winds occur during northers and the stunted forest trees on hilltops and peaks, particularly at high altitudes, is partially a result of exposure to wind. In some places on the volcano crater walls vegetation has not been able to establish itself mainly because of steep gradient and soil wash. In general, an interaction of factors involving slope, climate and soil cause a variability in structure of the forests constituting a distinct characteristic in many sections of the Sierra. The range's climate during the northern winter does not have any pronounced effect on vegetation besides tree stunting, but storms and temperatures which occasionally are about 0 degrees Centigrade at the highest elevations tend to cause a movement of some animals, most noticeably birds, to lower altitudes.

The comparatively stable conditions existing since Pleistocene times between the physical and biological environments are reflected in the distribution and abundance of birds and mammals today in undisturbed parts of the Sierra. The influence of physical factors maintained an essentially forest habitat before man's arrival to which the avian and mammalian faunas became well adapted. Chapter III and Appendix C give an idea of the relationships of mammals and birds to the various forest formations. In the comparatively homogeneous environment of the range's climax rain forest the rather sparse distribution and lower abundance of many forms are evident in comparison with nonforest habitats. Allee (1907:461) commented on the decrease in abundance

139

of rain forest animals in Panama in places where there is the least light. This also appears to be true in the Sierra rain forest, although perhaps in this subformation light conditions are more variable and such a distribution less noticeable than in Panama.

Periodic local destruction of forests by natural means (volcanism, wind, fire, tree decay) caused varying alterations in bird and animal populations before man's appearance. Forest regeneration that followed was characterized by areally small seral stages in the extensive climax forests. Although some species, such as the Howler Monkey, Great Tinamou and Great Curassow, requiring undisturbed, essentially climax forest habitat, were excluded, many other species adapted themselves to these habitat modifications. Relatively few nonforest birds and mammals, however, were able to enter the mountains. One can speculate about these nonforest elements and their ecological status in post-Pleistocene times. Birds associated primarily with unforested areas are now so numerous in species and numbers that perhaps some were originally forest species gradually adjusting to other habitats as the latter increased in area. Between Pleistocene and human times, therefore, there was a more or less continual, but minor, fluctuation in bird and mammal distribution and abundance taking place in the Sierra under natural controls acting principally through vegetation. Since the region is now about half forested with much of this being climax growth, physical factors are still operative in undisturbed areas. Their effects in the other sections of the Sierra have been changed by the landscape modifications brought about by man.

Human effects on the flora and fauna. The most profound change in the Sierra de Tuxtla by the human element has occurred in the natural vegetation. Man's influence on the plant cover of the large forested part of the region is now negligible or nonexistent and has likely been so since humans first arrived. The vegetation of the remaining area, however, has been destroyed or significantly altered by cutting and fire for many centuries. Small villages and the surrounding subsistence agricultural plots of the Sierra's earliest inhabitants were only minor interruptions in the essentially unbroken original vegetation cover. In these early times the human population was so sparse that the slash-burn agriculture practiced by most people affected relatively small areas. The fairly dense population and larger settlements of

140

later preconquest centuries, mostly on the Sierra's southern slopes, greatly extended the removal of vegetation. Since extensive sections of the range have more fertile soils than most tropical regions, however, I would judge that there was less migratory tillage and reclamation of slash-burned land by natural vegetation during the Sierra's early history than in regions where poorer soils occur. The range's more productive soils permitted the cycle of slash-burn cultivation to be longer and require less frequent abandonment and shifting of people to new land. Of course, this situation also depends on population density in any region considered, but the fairly dense population in the Sierra would have also tended to retard original vegetation regeneration. In spite of the expanding preconquest population and the clearing of new land, this factor of less shifting in agricultural activity is probably one of several reasons why almost half of the Sierra is still essentially undisturbed by human activity.

After the marked decrease in population subsequent to the conquest, since the 18th century there has been a more or less continual population rise evidenced in the Sierra by settlement expansion coincident with slash-burn cultivation and a great increase in permanent subsistence agriculture (Plate XXXI). Thus, in the settled areas natural forest vegetation now occurs mainly in scattered remnants, mostly composed of secondary stages. Where land is not under cultivation, there is a proliferation of weedy fields mixed with low, dense thickets and occasional woodland patches. Much of this low vegetation is either cut or burned before it is able to reach even an initial forest stage. The forest remnants and edges of the continuous forest are consistently exploited for firewood. In and near villages and larger urban centers there is a high population of domestic animals, such as burros, cattle, pigs and chickens, compacting the soil, altering vegetation and destroying plants to a diminishing degree radially from the center. The barren zone surrounding towns in Central America, referred to by Cook (1909:9; 1919:314-315), is not usually apparent around the Sierra's urban localities. Although vegetation is reduced or sparse and some bare or grass areas and a maze of well worn trails exist, there is no complete denudation of plant cover. Around some villages and larger towns there is a permanent type of horticulture (Bartlett, 1956:695-696). The natural vegetation has been replaced by groves of fruit trees and bananas with plots of vegetables scattered here and

141

Plate XXXI. Cultural Landscape

2.5 km. west San Andrés Tuxtla, 300 m. a.s.l. A typical view in a long-settled part of the range shows secondary forest remnants, thickets, newly burned fields and poor pastures. On the steep hill slope at upper right is a small cluster of huts (light area) with well worn trails leading to overgrazed areas where some erosion is evident. Mango trees, as in the foreground, have been widely planted for fruit.

142

there. Tapalapan, Mecayapan and some of the villages about Lago Catemaco are examples of such places. In many of the cultivated and partly open sections the people have planted fruit trees and also live fences of trees as field borders, such vegetation forming a conspicuous feature of the cultural landscape.

The effects of slash-burn activities in relation to destruction and regrowth of the Sierra's natural vegetation vary in different parts of the range. Where annual rainfall is less and the dry season more pronounced on the southern side, plant destruction by fire is facilitated, natural vegetation, especially woody plants, is able to regenerate less rapidly, and grasses and weeds are able to expand their coverage. On the less fertile soils on the south side of the Volcán Santa Marta massif these processes are particularly evident where the original growth was initially sparse. The intensive cultivation in the permanent agriculture following the slash-burn in many places has served further to limit the natural plant growth to hedgerows and isolated patches of secondary growth. Slash-burn operations, which are practiced today at and near the edges of the continuous humid forest and in scattered locations where small tracts and remnants of semideciduous forest types exist, are generally effective in killing almost all natural plant material above ground. The state of complete denudation and exhaustion of the soil, mentioned by Cook (1919:310) in describing the *milpa* system, is not reached in many places in the range. Even though there is no reforestation, the abundant rainfall, friable and fertile volcanic soils, rapid growth of plants supplying nutrients and general absence of destructive erosion, are conducive to more prolonged cultivation, more rapid soil recovery and in places more or less permanent land use. Regeneration of natural vegetation in the slash-burn area is especially effective in localities remote from villages in the zone of higher rainfall and less weathered soils occupied by much of the virgin rain forest.

Slash-burn operations at the edges of and within the continuous humid tropical forest create conditions whereby strong sunlight and heavy rain are able to affect soils and plant growth in the clearing and at the forest edge. A consequence has been the greatly increased extent of forest edge vegetation that is particularly noticeable and is an important vegetation modification in the Sierra. When downed forest trees and limbs are burned the adjacent undisturbed forest occasionally ignites. This usually results only in the destruction of vegetation at

143

the extreme edge which has been subject to some drying from exposure to sun. Watters (1960:82) mentions the effects of burning and dry conditions on soil composition and structure, and consequent favoring of more xerophytic and pyrophyllous plants. Although I observed the greater tendency to burn of forest edge vegetation, I did not notice that it showed a more deciduous character in humid rain forest areas. When clearings are kept under cultivation and mostly free of new natural vegetation, a very dense growth of herbaceous plants, shrubs, saplings, and vines, often differing in species from those in the forest, rapidly forms a sometimes almost impenetrable wall at the forest edge. On many occasions I have had considerable difficulty in entering the forest because of this thick vegetation. If the clearing is abandoned and regeneration commences, plants are aided in their spread from the forest edge by generation from this already tall growth. Although exposure facilitates the formation of edge vegetation, it often causes the solitary large trees left standing after slash-burn activities (Plate XXIII) to deteriorate and die. The epiphytes and climbers on their trunks and limbs also are adversely affected from exposure.

Slash-burn cultivation in the Sierra de Tuxtla is essentially similar in process to that employed in other regions of Latin America and the Old World. As in, for example, the bush fallow and kaingin agriculture of Nigeria and the Philippines, most tree roots and stumps are not removed, leaving sources for regeneration of the natural vegetation (Stamp, 1938:36; Pendleton, 1942:195). The shifting cultivation within the edges of the Sierra's rain forest often results in noncontiguous clearings so new growth quickly takes over upon abandonment and grasses are not able to dominate. This is in contrast to the gradual cultivation from the large cleared areas of permanent agriculture. Here the slash-burn activity is gradually pushing up the mountain slopes followed by establishment of grass and weed fields. A similar contrast is apparent on the mountains of Mindanao in the Philippines (Pendleton, 1942:207). There is not any noticeable effort by the Sierra's inhabitants to aid in the restoration of soil fertility to the slash-burn areas by crop variety and rotation, cover crops, or, for example, as in southern Nigeria (Stamp, 1938:36), by a fallow system. In Nigeria this bush fallowing is actually a system of permanent cultivation.

The general cultural processes by which man has changed the vegetation of the Sierra have involved a) initially a gradually expanding

144

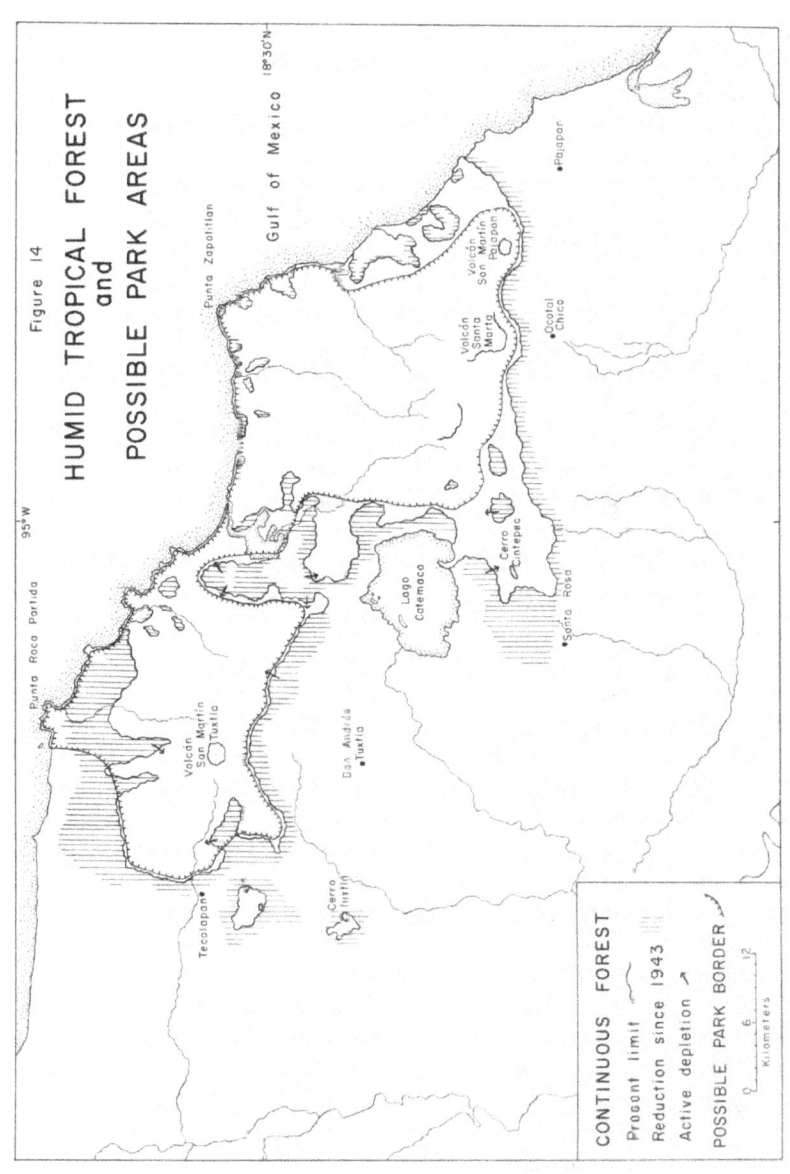

Figure 14

HUMID TROPICAL FOREST
and
POSSIBLE PARK AREAS

Gulf of Mexico

CONTINUOUS FOREST

Present limit
Reduction since 1943
Active depletion

POSSIBLE PARK BORDER

0   6   12
Kilometers

scale of plant cover destruction and change by shifting cultivation, and b) later a transformation of this type into predominantly sedentary permanent agriculture over much of the southern side of the range. Whetten (1948:575) listed 31.4 per cent of Veracruz as forested. Figure 7 shows the much higher proportion (almost 50 per cent) of the Sierra that is forest covered. Figure 14 shows that the reduction in area of the humid forest in about two decades by slash-burn cultivation has been considerable. Not portrayed in this figure are the many small remnants of other forest formations reduced in size by this method of land use. Aerial photographs of the southern slopes in places show a mosaic pattern of slash-burn clearings with a density often resulting in reduction of natural vegetation to relatively narrow borders between them. I noted some areas, several rather extensive, that had been cleared within the last decade and apparently had never been used for agriculture. I was informed that the recent clearing of the primary forest above Colonia Huatusco was for establishing a coffee plantation, but those involved did not know how to carry out the project and it was abandoned. A few small cattle herds were pastured in this area.

The rain forests of the Sierra are a unique formation whose components are closely adapted to the forest association. The juxtaposition of the tops of the upper story and the intermixture of branches and lianas in the canopy form a complex undergoing severe disruptions if a tree is felled by human action or natural means. This is occurring in the two lumbering operations I observed in the range. At the present time, however, these activities are not causing major forest depletion. I saw no evidence of clear cutting and there seemed to be a general policy of selectivity. The multiplicity of tree species, the general absence of commercially valuable forms and the operational difficulties involved, make large scale efforts expensive and largely impracticable. The effects of charcoal making on the vegetation of the Sierra are also not of great importance, being extremely limited in extent and frequency. The clearing of forests by slash-burn methods remains the major factor affecting natural vegetation in the Sierra de Tuxtla. This practice is now subject to some regulation under the forest law, permitting cutting only on slopes with less than 15 per cent grade. The control of this human influence on the flora, however, requires much closer observation and enforcement.

Human alteration and removal of natural vegetation in the Sierra have exposed extensive areas to the physical environmental influences continually affecting the land. Although a protective cover of plant growth rapidly forms in many places, the exposure of the land surface, formerly shielded by forest, to the sunlight results in soil desiccation and loosening which is conducive to an increased rate of erosion. The heavy precipitation during the rainy season, no longer broken by the closed canopy of forest, is more easily able to contact the ground surface and cause soil wash, occasional slumping and some minor gullying. Accelerated runoff on land surfaces and in streams is particularly evident in the pine-oak areas modified by man where natural vegetation is much less dense than in places where the humid tropical evergreen forest occurs. Although all of the Sierra's streams carry some sediment during rainy periods, those on the southern slopes in the cleared areas show greater turbidity than do those flowing largely in humid forest. The continual removal of forest vegetation by man will undoubtedly increase this runoff and add to soil loss.

Man had small effect on the Sierra's avian and mammalian populations in the earliest centuries. In spite of his greater dependence then on hunting for subsistence, his low numbers and less efficient weapons precluded any but very minor reductions in animal numbers. He was able to decrease animal populations only locally near villages or cultivated land. As the human population of the Sierra increased during the centuries before the conquest, there was greater pressure from hunting and a consequent areal expansion in game pursuit having a more widespread effect in reducing total populations. During these times bows and arrows, slings, snares and traps were used, occasionally in communal hunts. There is no evidence when the dog was first employed in hunting, but it very likely was used often, especially for the terrestrial burrowing animals. Dogs are frequently used now, and are considered essential by some people for hunting the Red Brocket and for various rodents.

Influence on animal populations from hunting probably declined somewhat after the conquest because of the decrease in human population. The acquisition of firearms, however, enabled hunters to procure birds and mammals more easily for food. There is little historical information on hunting in the Sierra, but it is probable that very few people were able to afford firearms until the modern period. As was mentioned previously, even today not many people can afford to purchase a gun

147

or buy the ammunition to hunt steadily. Nevertheless, the efficiency of firearms has contributed to the reduction of animals, particularly the larger and more gregarious mammals such as the monkeys, and the solitary mammals such as the Paca and Agouti. Hunting with firearms has not only markedly reduced some species of game birds and mammals in certain localities, but caused a decrease generally in the mountain range. A recent example of this decrease is the White-collared Peccary, which occurred in large numbers as recently as 25 years ago, but has now almost completely disappeared. Some other species noticeably decreasing in abundance in many localities due to hunting are the Great Curassow, Crested Guan, Baird's Tapir, Paca and Red Brocket. The combination of more and better weapons and a greatly increased and more widely settled human population have resulted in greater hunting pressure than formerly on many animals in spite of the increasing proportion of people engaged in agriculture and urban occupations.

A more recent development is the ability of some urban residents to hunt in the Sierra during leisure time for sport. Argudin (1955:33) told of a small local hunting party northeast of Volcán San Martín Tuxtla taking a Baird's Tapir, 24 curassows (possibly including guans), two Pacas, two Red Brockets, two Kinkajous and one "tigre serrano." In view of the changing status of many people in the country, this type of hunting is likely to have an increasing effect on the fauna.

Man's effect on the avifauna in regard to hunting is not easily ascertained with the exception of several larger species. My observations showed that most local hunters are more selective in taking birds than they are mammals, and will more often try to secure those yielding the most meat or those they consider more palatable. As with mammals, though, few opportunities to secure meat for consumption are let pass. At irregular intervals in various Sierra localities I saw people shoot grebes, ducks, herons, kites and toucans, some of this being done by nonresidents. Total populations of most bird species, however, have not been significantly affected by hunting.

The composition and total populations of birds and mammals in the Sierra have been changed significantly by man indirectly by removal and alteration of the natural vegetation. The humid tropical forest is able to support a comparatively low population density of the larger mammals and birds, such as the Howler Monkey, the cats, Baird's Tapir and the Great Curassow. Most of these animals will not remain close to human habitations. Consequently, continual loss of forest habitat and establishment of new settlements at the edges and within the forest has caused a significant decline in total Sierra populations of these forest dwelling mammals and birds. Some smaller passerine birds of the humid primary forest are able to adapt easily to various stages of secondary growth. Frequent and sometimes uncontrolled use of fire in slash-burn operations and also in grass and cultivated areas in the range not only contributes to loss of animal habitat and food supplies, but destroys some of the small vertebrate and invertebrate forms upon which larger mammals and birds depend for food. As the region's forest cover is reduced the introduction and dispersal of animals inhabiting forest edge, secondary growth, thickets and open country increases. Although no direct studies have been made concerning this change, it is probable that there has been an increase in several bat species and a rise in numbers of opossums, armadillos, raccoons, rabbits, some small rodents, and possibly weasels. With some of these forms an over-all increase has definitely taken place in recent years. The avifauna is undergoing a similar change in abundance and composition from man's influence on the vegetation. The large number and variety of nonforest dwelling bird species (Appendix C, habitats B, C, E) and the movements of some into the Sierra can be observed during comparatively brief observations. A few of the most noticeable of these avian forms are Groove-billed Ani, Vermilion-crowned Flycatcher, Brown Jay, Boat-tailed Grackle, Red-eyed Cowbird, Yellow-faced Grassquit, Blue-black Grassquit and White-collared Seedeater. Some larger open country bird species, such as the White-tailed kite, Crested Caracara, Laughing Falcon and Roadside Hawk, have also increased in frequency of occurrence.

5.2   Prospect

Problems in the region. The Sierra de Tuxtla possesses major problems basically related to four factors - location, topography, climate and soils. The range is located on the Gulf coastal plain in southeastern Mexico directly on the route from the populous highlands via Veracruz city to Tabasco and the Yucatan peninsula. The major coastal road completing the connection of the highlands with the southeastern areas has recently been completed; the section, partly in the Sierra, between Catemaco and Acayucan has only been paved for slightly more than a decade. Thus, with the major transportation route of southeastern Mexico crossing the range, the area's future is directly involved with the economic development taking place along this route. At the present time within the Sierra this development consists principally in agricultural and livestock activities plus urban commercial enterprises. Topographic conditions are such that, despite many steep slopes, deep ravines and gorges, a large amount of the Sierra's land area has a moderate to slight gradient. This condition, with the fact that the region's soils are generally more fertile than most tropical soils, is conducive to a continuous and expanding use of land for agriculture. Finally, the mild climate and abundant rainfall provide good growing conditions for crops and thereby make the Sierra an attractive location for new settlements.

Intimately related to the foregoing factors are the two most important problems: a) the rapid population increase and b) the use of land in the Sierra. Table VII shows that total population in the last thirty years has increased by 40 per cent per decade. This represents a significant immigration plus the normal high increase in the resident population. The impact of this increase on the original vegetation is causing the landscape in the Sierra to change from a forest to an agricultural one. If this rate continues, the pressure to expand agricultural and grazing areas will undoubtedly be greater. The question arises as to what effects such expansion has already had, and will have, if continued, on the soils, water resources and plant and animal life in the Sierra. That the latter two elements have been greatly altered with the concomitant reduction in abundance of many native species, there is ample evidence.

It is more difficult, however, to determine the general effects on soil and water resources in the watershed of the Sierra de Tuxtla. It is apparent that moderate to slight erosion has occurred and is continuing in the unforested parts, particularly on some steep slopes where annual burning and grazing is carried out. This will accelerate with a progressive reduction in soil fertility under present agricultural methods, especially if the steeper slopes and higher lands continue to be converted to agricultural use; accelerated erosion will occur on such terrain despite the rapid growth of vegetation which is now helping to prevent severe erosion in the less steep lands cleared of forest at lower elevations. In relation to agriculture at higher altitudes on steep slopes, Gourou (1956:344) said that to be successful in supporting increasing populations intensive must replace extensive production of food products in tropical cultivation. To accomplish this he indicated that an inversion of cultivated areas must take place utilizing lower topographic levels that are usually the best for agriculture. The poor ability of maize to inhibit erosion will be a factor in the Sierra in view of its extensive planting and the lack of agricultural conservation practices involving crop rotation, contour plowing and soil management. I believe the entire removal of the great forests in the Sierra will sufficiently disturb the ecosystem so that generally adverse effects will occur on soils, water table and flow and agricultural potential of its land; these effects can only be partially compensated for by conservation procedures, which are at present almost nonexistent. The Secretaria de Agricultura y Ganadería is making efforts to improve the situation by demonstrations and education, but much remains to be accomplished, particularly in view of the rapid colonization of virgin lands and the remoteness of many parts of the range.

The Sierra de Tuxtla is essentially well-watered with Lago Catemaco and many streams providing sufficient amounts for the existing population as well as for hydroelectric power. The forests now act as a conservator for the water supplies of the Sierra. With continued forest removal, eventual permanent cultivation and lack of soil conservation, leaching and precipitation runoff will increase and the total water table may be lowered. Although humidity is high, the severity of the dry season, evidenced now by reduction of stream water supply and desiccation of unvegetated soil, especially in the less humid areas, will also probably be increased by forest denudation (cf. Thornthwaite, 1956:578, 582).

Because a large proportion of the region consists of land having moderate or slight gradient, it is possible eventually to bring most of it under cultivation and to introduce grazing or agriculture on the steeper sections at higher elevations. This may occur in the future if the present trends continue, but in my estimation such a condition would not be best for the Sierra's land considering the physical environmental factors continually acting upon it. The utilization of the steeper slopes for tree crops rather than cultivation or grazing would be a better use of

151

the higher land, if it is considered necessary to use the entire Sierra for subsistence and commercial agriculture (cf. Cook, 1919:325). Scientific silviculture in this region where trees grow exceptionally well would contribute to the economy as well as assist in watershed protection. In the final analysis, whether the Sierra is eventually employed almost entirely as an agricultural and grazing area or whether a part will be left in forest for other purposes, rests on the values and intentions of those in a position to exercise control. The problems of population increase and land use require strong action if they are to be solved for the long-term benefit of both the people and the land. To accomplish this it is necessary for those in government concerned with resource development and conservation to formulate and implement policies based on an evaluation of the Sierra's total potential in terms of natural resources, economic development and possibilities for recreational use. This approach, to be efficacious, must embody considerations based not only on sound land use but upon conservation principles.

Conservation. Unfortunately, the general conservation outlook in the Sierra de Tuxtla is not good. Slash-burn agriculture on new lands is at present the only means practicable for part of the expanding population to subsist. This is impressed upon an observer as when, for example, in May 1962, I counted 15 fires burning on the forested slopes above the Bahía Sontecomapan, and the atmosphere over parts of the Sierra was hazy with smoke. The destruction of the forests is carried out, especially in remote areas, despite the excellent forest law. Article 126 of the regulations of the Forest Law (Secretaria de Agricultura y Ganadería, 1961:63) states that the Secretary of Agriculture and Livestock can authorize the cutting of forests for opening new lands for agriculture only when the slope of the land is less than 15 per cent. In some sections I observed recent cutting on considerably steeper slopes. The chief of the forest sector, working mainly alone in this large area and lacking adequate transportation facilities, is handicapped in reconnaissance and law enforcement.

The conservation picture in relation to the fauna is even less promising at the present time. The taking of permanently protected species such as the monkeys, tapir and various birds continues. Since there are no wildlife officers, enforcement of the fine game law is not possible. The chief of the forest sector is empowered to enforce this law, but is not even able to do so adequately with respect to the forest law. Many people outside of the larger towns did not appear to be cognizant of laws concerning forests or game. Yet there were some who deplored the cutting and burning of the forests and seemed to be aware of the consequences to the land and wildlife. A few land owners expressed their concern over this situation and indicated their intention to take measures on their properties to guard against erosion and indiscriminate forest destruction. The loss of habitat by permanent destruction of the original and secondary vegetation is a major cause of faunal impoverishment in the Sierra. Some residents are aware of present conditions and the need for proper land use and conservation in the region. It is extremely difficult, however, to reach many parts of the Sierra either with education or law enforcement in this regard. It is likewise difficult to introduce new agricultural and soil conservation practices because many of the rural people possess ingrained traditions with respect to use of the land; in some cases apathy or even antagonism is shown toward the introduction of new methods of land use, though the people would probably benefit in the future as a result of adopting them.

5.3   Summary

The Sierra de Tuxtla is a discrete topographic uplift possessing distinct physical, biological and cultural characteristics. Its volcanic composition forms a geological and geomorphological basis upon which its plant cover and elements of its rich avian and mammalian faunas have been affected by physical factors and the activities of the human population.

For the hundred million or more years of its existence elements of the physical environment in the form of water, wind and volcanic disturbances have acted upon the Sierra dynamically and in effect have molded its form into what is evident today. The flora and fauna upon its surface have reacted to these influences with a series of changes that likewise continue unceasingly. The several thousand years of man's occupance are only a minute fraction of the range's existence in geological time, not sufficient for much except relatively minor effects upon its physical surface, principally in the form of erosion. Man's impact upon the flora and fauna, however, has been considerable over much of the region. Such comparatively recent changes can often be observed and analyzed more clearly than can those taking place during the Sierra's existence through past geological periods.

The surface of the Sierra de Tuxtla, though plainly showing in basic form the cratered cones and lava ridges of its volcanic origin, also possesses the rounded contours and radial ravine patterns resulting from centuries of action by mechanical and chemical weathering in a tropical climate. The relatively stable balance existing between the land surface features and the almost continuous plant cover during prehuman times is now radically changed over about half of the region. Vegetation destruction by cutting and fire, the introduction of new plants, the utilization of land for an agricultural and grazing economy, have modified much of the Sierra's landscape. With the extensive loss of vegetative cover and the resulting greater exposure of the land surface, particularly on the already less humid southern slopes of the Sierra, soil structure has been changed, erosion and water runoff increased so that it is possible a general trend toward drier conditions exists.

The continuing modifications in the nature and extent of the plant cover are causing more changes in bird and mammal populations. The shift from forest dwelling to nonforest dwelling forms is progressing, with some species of the former approaching local extinction. Use by the human population as a source of food or for sale as pets is contributing to the decline of some forms.

A high rate of human population increase plus immigration has resulted in an expansion of settlement and agricultural land use. In addition to its aforementioned effects on plant and animal life, this expansion has resulted in damage to existing forests by fire and an increase in domestic animals with detriment to natural vegetation where they roam uncontrolled. Although attempts are being made to improve agriculture and grazing, at present there is little external evidence of this. The outlook in regard to forest and wildlife conservation is poor. Enforcement of laws pertaining thereto is a major problem involving lack of personnel and transportation facilities, difficulty of access to certain areas, and the attitudes of the human residents.

The future of the Sierra de Tuxtla lies in a balanced program embodying education, economic development and strong conservation measures. If the Sierra is to conserve its resources, increase its productivity and preserve its natural beauty, there must be a concerted effort to establish working man-land relationships based on a concept of the Sierra as a distinct physical and biological entity. Effective steps should be taken to improve agriculture, protect forests, waters and wildlife and impress upon its people the necessity of having a long-range view in regard to proper land use and conservation.

BIBLIOGRAPHY

Allee, W. C.
    1926. Distribution of animals in a tropical rain-forest with rela-
    tion to environmental factors. Ecology 7:445-468.

Alvarez del Toro, M.
    1952. Los animales silvestres de Chiapas. Tuxtla Gutierrez,
    Chiapas. 247 p.

Amadon, D. and D. R. Eckelberry
    1955. Observations on Mexican birds. Condor 57:65-80.

Argudin, R. S.
    1955. Relatos de cacerías en Los Tuxtlas, Veracruz. Talleres de
    Morales. Hermanos Impresores, S. A. México, D.F. 259 p.

———.
    1962. Los taladores. In "Los Tuxtlas," Revista regional. San
    Andrés Tuxtla, Veracruz. February, No. 6, p.10-11.

Barberena Vega, M.
    1962. Información general del Estado de Veracruz. 2 vols. Inst.
    de Ciencias, Univ. Veracruzana, Jalapa.

Bartlett, H. H.
    1956. Fire, primitive agriculture, and grazing in the tropics. In
    "Man's Role in Changing the Face of the Earth," W. L. Thomas,
    Jr., ed., Part II, p. 692-720. Univ. of Chicago Press, Chicago.

Beard, J. S.
    1944a. Climax vegetation in tropical America. Ecology 25:127-158.

———.

    1944b. The natural vegetation of the island of Tobago, British West Indies. Ecol. Mono. 14:135-163.

———.

    1953. The savanna vegetation of northern tropical America. Ecol. Mono. 23:149-215.

———.

    1955. The classification of tropical American vegetation types. Ecology 36:89-100.

Beltrán, E.
    1946. Los recursos naturales de México y su conservación. Sec. Educ. Pública, La Biblioteca Enciclopédica Popular 106. México, D.F. 96 p.

———.

    1961. Temas forestales, 1946-1960. Talleres de la Editorial Cultura, T.G., S.A. México, D.F. 283 p.

Blom, F., and O. La Farge
    1926, 1927. Tribes and temples, a record of the Tulane University expedition to Middle America. 2 vols. Tulane Univ. of Louisiana, New Orleans. 536 p.

Boettiger de Alvarez, M.
    1961a. Montepío, Ver. In "Revista Jarocha," Tacubaya, D.F., February, No. 11, p. 15-16, 61.

———.

    1961b. Pinceladas descriptivas de Sontecomapan, Ver. In "Revista Jarocha," Tacubaya, D.F., February, No. 11, p. 17, 61.

Borah, W., and S. F. Cook
    1963. The aboriginal population of central Mexico on the eve of the Spanish conquest. Ibero-Americana 45, 157 p.

Brands, G. J.
    1944. Meteorology. McGraw-Hill Book Co., Inc. New York. 235 p.

Brodkorb, P.
    1943. Birds from the Gulf lowlands of southern Mexico. Univ. Michigan Museum Zool. Misc. Publ. 55, 88 p.

Brown, W. H.
    1919. Vegetation of Philippine mountains. Bur. of Printing, Manila. 434 p.

Brown, W. H., and A. S. Argüelles
    1917. The composition and moisture content of the soils in the types of vegetation at different elevations on Mount Maquiling. The Philippine J. Sci. 12:221-233.

Budowski, G.
    1959. The ecological status of fire in tropical American lowlands. In "Actas del XXXIII Congr. Intern. de Americanistas" 1:264-278. San Jose, Costa Rica.

Burkart, H. B. (Direktor)
    1835. Über die Ausbrüche des Jorullo und des Tustla. Neues Jahrb. für Minerologie, Geognosie, Geol. und Petrefaktenkunde, p. 36-39.

Burt, W. H.
    1949. Present distribution and affinities of Mexican mammals. Ann. Assn. Am. Geogr. 39:211-218.

Carabia, J. P.
     1945. The vegetation of Sierra de Nipe, Cuba. Ecol. Mono.
     15:321-341.

Cline, I. M.
     1926. Tropical cyclones. The Macmillan Co. New York. 301 p.

Comisión del Papaloapan
     1958. Datos hidrológicos de la cuenca del Río Papaloapan. Dir.
     Técnica - Oficina de Hidrometeorología, Bol. Hidrológico 10,
     597 p.

Contreras Arias, A.
     1942. Mapa de las provincias climatológicas de la República
     Mexicana. Sec. de Agr. y Fomento, Dir. de Geogr., Meteorol. y
     Hidrol. Inst. Geogr. México, D.F. 54 p.

Cook, O. F.
     1909. Vegetation affected by agriculture in Central America.
     United States Dept. Agr., Bur. Plant Industry Bull. 145, 23 p.

————.

     1919. Milpa agriculture, a primitive tropical system. Ann. Rep.
     Smiths. Inst., p. 307-326, pl. 1-15.

Cook, S. F., and L. B. Simpson
     1948. The population of central Mexico in the sixteenth century.
     Ibero-Americana 31, 241 p.

Covarrubias, M.
     1946. Mexico south: the isthmus of Tehuantepec. A.A. Knopf,
     New York. 427 p.

Cushing, S. W.
     1921. The distribution of population in Mexico. Geogr. Rev.
     11: 227-242.

Dalquest, W. W.
    1949. The White-lipped Peccary in the state of Veracruz, Mexico.
    An. Inst. Biol. 20:1-3.

Darlington, P. J., Jr.
    1957. Zoogeography: the geographical distribution of animals.
    John Wiley and Sons, Inc., New York. 675 p.

Davis, L. I.
    1952. Tropical woods. Sixteenth Breeding Bird Census, Audubon
    Field Notes 6:314-315.

————.
    1955. Tropical rain forest. Nineteenth Breeding Bird Census,
    Audubon Field Notes 9:425-426.

Davis, L. I., and J. Morony, Jr.
    1953. Lowland tropical forest. Seventeenth Breeding Bird Cen-
    sus, Audubon Field Notes 7:325-353.

Davis, T. A. W., and P. W. Richards
    1933. The vegetation of Moraballi Creek, British Guiana: an
    ecological study of a limited area of tropical rain forest. J. of
    Ecol. 21:350-384.

————.
    1934. Ibid., 22:106-155.

Díaz del Castillo, B.
    1908-1916. The true history of the conquest of New Spain. G.
    Garcia, ed. and publ., transl. by A. P. Maudslay. 4 vols. Bed-
    ford Press, London.

Dirección General de Geografía y Meteorología
    1962. Carta del tiempo. January-December, 1962. Servicio
    Meteorológico Mexicano. México, D.F.

Doerr, A., and L. Freile
    1956. Population distribution in Mexico - 1950. J. of Geogr. 55:235-242.

Dorf, E.
    1959. Climatic changes of the past and present. Contrib. Museum Paleontol. Univ. Michigan 13:181-210.

Drucker, P.
    1943. Ceramic sequences at Tres Zapotes, Veracruz, Mexico. Smiths. Inst. Bur. Am. Ethnol. Bull. 140, 148 p.

Drucker, P., R. F. Heizer, and R. J. Squier
    1959. Excavations at La Venta, Tabasco, 1955. Smiths. Inst. Bur. Am. Ethnol. Bull. 170, 312 p.

Dunn, G. E., W. R. Davis, and P. L. Moore
    1955. Hurricanes of 1955. United States Dept. Commerce, Weather Bur., Monthly Weather Rev. 83:315-326.

————-.

    1956. Hurricane season of 1956. United States Dept. Commerce, Weather Bur., Monthly Weather Rev. 84:436-443.

Edwards, E. P., and R. E. Tashian
    1959. Avifauna of the Catemaco basin of southern Veracruz, Mexico. Condor 61:325-337.

Eisenmann, E.
    1955. The species of Middle American birds. Trans. Linn. Soc. New York. Vol. 7, 128 p.

Epstein, M., ed.
    1933. Mexico. In "The Statesman's Year-Book, Statistical and Historical Annual of the States of the World for the Year 1933." Macmillan and Co. Ltd. London. 1471 p.

———.

1943. Mexico. In "The Statesman's Year-Book, Statistical and Historical Annual of the States of the World for the Year 1943." The Macmillan Co., New York. 1469 p.

Escobar, M. de
1794. Report in "Archivo General de la Nación, Ramo Industria y Comercio," México, D.F., vol. 31, proc. 24.

Fassig, O. L.
1913. Hurricanes of the West Indies. United States Dept. Agr. Weather Bur., Gov. Printing Office, Washington. 28 p., 25 pls.

Firschein, I. L.
1950. A new toad from Mexico with a redefinition of the *cristatus* group. Copeia No 2:81-87.

Firschein, I. L., and H. M. Smith
1956. A new fringe-limbed Hyla (Amphibia:Anura) from a new faunal district of Mexico. Herpetologica 12:17-21.

Fosberg, F. R., B. J. Garnier, and A. W. Küchler
1961. Delimitation of the humid tropics. Geogr. Rev. 51:333-347.

Foster, G. M., Jr.
1940. Notes on the Popoluca of Veracruz. Inst. Panam. de Geogr. e Hist. Publ. Num. 51, 41 p.

———.

1942. A primitive Mexican economy; an ethnographical study of the Popoluca Indians of Veracruz. Mono. Am. Ethnol. Soc. No. V, 115 p.

Friedlaender, I.
1923. Uber das Vulkangebiet von San Martin Tuxtla in Mexiko. Zeitschr. für Vulkanologie 7:162-187, pl. 17-34.

163

Friedmann, H., L. Griscom, and R. T. Moore
1950. Distributional check-list of the birds of Mexico. Part I. Pacific Coast Avifauna 29, 202 p.

Garbell, M. A.
1947. Tropical and equatorial meteorology. Pitman Publ. Co., New York. 237 p.

García, J. A.
1835. Eruptionen des Vulkanes von Tustla in den Jahren 1664 und 1793. Neues Jahrb. für Minerologie, Geognosie, Geol., und Petrefaktenkunde, p. 40-45.

Garriott, E. B.
1900. West Indian hurricanes. United States Dept. Agr. Weather Bur. Bull. H, 69 p.

Gobierno del Estado de Veracruz
1962. Datos preliminares del censo de población de 1960. Veracruz.

Goldman, E. A.
1920. The mammals of Panama. Smiths. Misc. Coll. 69, 309 p.

————.

1951. Biological investigations in Mexico. Smiths. Misc. Coll. 115, 476 p.

Gomez Mayorga, N.
1961. Vislumbre por la región de Los Tuxtlas. In "Revista Jarocha," Tacubaya, D.F., February No. 11, p. 9-11.

Goodnight, C. J., and M. L. Goodnight
1954. The Opilionid fauna of an isolated volcano in southeastern Veracruz. Trans. Am. Microscop. Soc. 73:344-350.

Gourou, P.

    1956. The quality of land use of tropical cultivators. In "Man's Role in Changing the Face of the Earth," W. L. Thomas, Jr., ed., Part I, p. 336-349. Univ. of Chicago Press, Chicago.

Griscom, L.

    1932. The distribution of bird-life in Guatemala. Bull. Am. Museum Nat. Hist. 64, 439 p.

————.

    1950 Distribution and origin of the birds of Mexico. Bull. Museum Comp. Zool. 103:341-382.

Hall, E. R., and K. R. Kelson

    1959. The mammals of North America. 2 vols. The Ronald Press Co., New York. 1083 p.

Holdridge, L. R.

    1945. A brief sketch of the flora of Hispaniola. In "Plants and Plant Science in Latin America," F. Verdoorn, ed., p. 76-78. Chronica Botanica Co., Waltham, Massachusetts.

Holmes, W. H.

    1907. On a nephrite statuette from San Andres Tuxtla, Vera Cruz. Am. Anthropol. 9:691-701.

Hurd, W. E.

    1929. Northers of the Gulf of Tehuantepec. United States Dept. Agr. Weather Bur., Monthly Weather Rev. 57:192-194.

Jaussoro, R.

    1961. El Cantón de los Tuxtlas (1872). In "Revista Jarocha," Tacubaya, D F., February, No. 11, p. 5-7.

Kenoyer, L. A.

    1929. General and successional ecology of the lower tropical rainforest at Barro Colorado Island, Panama. Ecology 10:201-222.

165

Kerber, E.

    1882. Eine alte mexikanischen Ruinenstaate bei San Andres Tuxtla. Verhandl. Berliner Gesell. für Anthropol., Ethnol. und Urgeschichte 14:448-489.

Leopold, A.S.

    1959. Wildlife of Mexico; the game birds and mammals. Univ. of California Press, Berkeley. 568 p.

Lowery, G. H., Jr., and W. W. Dalquest

    1951. Birds from the state of Veracruz, Mexico. Univ. of Kansas Publ. Museum Nat. Hist. 3:531-649.

Lowery, G. H., Jr., and R. J. Newman

    1949. New birds from the state of San Luis Potosi and the Tuxtla Mountains of Veracruz, Mexico. Occ. Papers Museum Zool. Louisiana State Univ. 22:1-10.

Lundell, C. L.

    1937. The vegetation of Peten. Carnegie Inst. Washington Publ. 478, 244 p.

Macías Villada, M.

    1960. Suelos de la República de México. Ing. Hidrául. en México 14:51-70, 63-73, 99-112.

Martin, P. S.

    1955. Distribution of vertebrates in a Mexican cloud forest. Am. Naturalist 89:347-361.

Martin, P. S., and B. E. Harrell

    1956. The Pleistocene history of temperate biotas in Mexico and eastern United States. Ecology 38:468-480.

Mayr, E.

    1946. History of the North American bird fauna. Wilson Bull. 58:3-41.

Medel y Alvarado, L.
1963. Historia de San Andrés Tuxtla, 1532-1950. Colección Suma Veracruzana, Serie Historiografía. 2 vols. Editorial Citlaltepetl, México, D.F., 555 p., 627 p.

Melgarejo Vivanco, J. L.
1960. Breve historia de Veracruz. Biblio. Facul. Filos. Letras, Univ. Veracruzana, Xalapa. 268 p.

Miller, A. H., H. Friedmann, L. Griscom, and R. T. Moore
1956. Distributional check-list of the birds of Mexico. Part II. Pacific Coast Avifauna 33, 435 p.

Miranda, F., and A. J. Sharp
1950. Vegetation in temperate regions of eastern Mexico. Ecology 31:313-333.

Mociño Suarez de Figueroa, J. M.
1870. Informe de Don José Moziño sobre la erupción del volcán de San Martín, Tuxtla (Vera Cruz) ocurrido en el año de 1793. Bol. Soc. Geogr. Estad. Rep. Mex. 2:62-70.

―――.

1874-1876. Descripción del volcán de Tuxtla por D. Joseph Mariano Moziño Suarez de Figueroa, botánico naturalista de la Real Expedición de Nueva España y de límites al norte de California. Año de 1793. La Naturaleza, Periodico Cient. Soc. Mex. Hist. Nat. 3:106-114.

Morley, S.G.
1956. The ancient Maya. 3rd. ed. Stanford Univ. Press, Stanford. 494 p.

Murray, G. E.
1961. Geology of the Atlantic and Gulf Coastal Province of North America. Harper and Brothers, New York. 692 p.

Nelson, E. W.

167

1897. Preliminary descriptions of new birds from Mexico and Guatemala in the collections of the United States Department of Agriculture. Auk 14:42-76.

Norton, G.
1951. Hurricanes of the 1950 season. United States Dept. Commerce, Weather Bur., Monthly Weather Rev. 79:8-15.

———.
1952. Hurricanes of 1951. United States Dept. Commerce, Weather Bur., Monthly Weather Rev. 80:1-9.

Ordoñez, E.
1941. Las provincias fisiográficas de México. Rev. Geográfico, del Inst. Panam. de Geogr. e Hist. 1:133-181.

Ortiz Monasterio, R.
1955-1957. Los recursos agrológicos de la República Mexicana. Ing. Hidrául. en México, vols. 9-11. (6 parts).

Page, J. L.
1930. Climate of Mexico. United States Dept. Agr. Weather Bur., Monthly Weather Rev., Supplement 33. United States Gov. Printing Office. Washington. 30 p.

Paso y Troncoso, F.
1905. Relación geográfica de la villa de Tustla, 1580. In "Papeles de Nueva España," 5:4-8. Madrid.

Paynter, R. A., Jr.
1955. The ornithogeography of the Yucatán peninsula. Peabody Museum Nat. Hist. Bull. 9, 347 p.

Pendleton, R. L.
1942. Land utilization and agriculture of Mindanao, Philippine Islands. Geogr. Rev. 32:180-210.

———.

1950. Agricultural and forest potentialities of the tropics. Agron. J. 42:115-123.

Richards, P. W.
1936. Ecological observations on the rain forests of Mount Dulit, Sarawak. J. of Ecol. 24:1-37; 340-360.

————.
1957. The tropical rain forest. Cambridge Univ. Press, Cambridge. 450 p.

Riehl, H.
1954. Tropical meteorology. McGraw-Hill Book Co., Inc., New York. 392 p.

Ríos Macbeth, F.
1952. Estudio geológico de la región de Los Tuxtlas. Asoc. Mex. Geol. Petrol. Bol. 4:325-376.

"Santiago Tuxtla, Rincón de Maravilla."
1961. Revista Jarocha. Tacubaya, D.F., February, No. 11, p. 12-14.

Schuchert, C.
1935. Historical geology of the Antillean-Caribbean region. John Wiley and Sons, Inc., New York. 811 p.

Sclater, P. L.
1857. List of additional species of Mexican birds, obtained by M. Auguste Sallé from the environs of Jalapa and San Andres Tuxtla. On a collection of birds received by M. Sallé from southern Mexico. Proc. Zool. Soc. London Part 25, p. 201-207, 226-230.

Sears, P. B.
>1952. Palyonology in southern North America I: archeological horizons in the basins of Mexico. Bull. Geol. Soc. Am. 63:241-254.

Secretaria de Agricultura y Ganadería
>1961. Ley forestal y reglamento de la ley forestal. Subsecretario de Recursos Forestales y de Caza, Edición del Inst. Nac. de Invest. Forest., México, D.F. 101 p.

Seifriz, W.
>1943. The plant life of Cuba. Ecol. Mono. 13:375-426.

Seler-Sachs, C.
>1922. Altertümer des Kanton Tuxtla im staate Veracruz. In "Festshrift Eduard Seler," p. 543-556. Verlag von Strecker und Schröder, Stuttgart.

Shelford, V. E.
>1941. List of reserves that may serve as nature sanctuaries of national and international importance, in Canada, the United States, and Mexico. Ecology 22:100-110.

Shreve, F.
>1914. A montane rain forest. Carnegie Inst. Washington Publ. 199, 110 p.

Soto, J.
>1961. Laguna de Catemaco (1851). In "Revista Jarocha," Tacubaya, D.F., February, No. 11, 18-19, 61.

Spinden, H. J.
>1924. The reduction of Mayan dates. Harvard Univ., Peabody Museum Papers, 6.

————.

>1928. The population of ancient America. Geogr. Rev. 18:641-660.

——————.

1943. Ancient civilizations of Mexico and Central America. 3rd. ed. Am. Museum Nat. Hist., New York. 271 p.

Stamp, L. D.
1938. Land utilization and soil erosion in Nigeria. Geogr. Rev. 28:32-45.

Standley, P. C.
1920-1926. Trees and shrubs of Mexico. Contrib. United States Nat. Herb. 23:1-1721 (five parts).

Standley, P.C., J. A. Steyermark and L. O. Williams
1946-1962. Flora of Guatemala. Chicago Nat. Hist. Museum. Fieldiana: Botany 24: parts I-VII.

Steinberg, S. H., ed.
1953. Mexico. In "The Statesman's Year-Book, Statistical and Historical Annual of the States of the World for the Year 1953." St. Martin's Press, Inc., New York. 1595 p.

——————.

1962. Mexico. In "The Statesman's Year-Book, Statistical and Historical Annual of the States of the World for the Year 1962-1963." St. Martin's Press, Inc., New York. 1702 p.

Stirling, M. W.
1943. Stone monuments of southern Mexico. Smiths. Inst. Bur. Am. Ethnol. Bull. 138, 84 p.

Sumichrast, F.
1869. The geographic distribution of the native birds of the department of Vera Cruz, with a list of the migratory species. Boston Soc. Nat. Hist. Mem. 1:542-563.

Sumner, H. C.

  1941a. Tropical disturbances of September, 1941. United States Dept. Commerce, Weather Bur., Monthly Weather Rev. 69:264-266.

————.

  1941b. North Atlantic tropical disturbances of 1941. United States Dept. Commerce, Weather Bur., Monthly Weather Rev. 69:363.

————.

  1944. North Atlantic hurricanes and tropical disturbances of 1944. United States Dept. Commerce, Weather Bur., Monthly Weather Rev. 72:237-240.

Tannehill, I. R.

  1938. Hurricanes, their nature and history. Princeton Univ. Press, Princeton, New Jersey. 257 p.

————.

  1950. The hurricane. United States Chamber of Commerce Misc. Publ. No. 197, 9 p. 16 figs.

Tannenbaum, F.

  1929. The Mexican agrarian revolution. The Macmillan Co., New York. 543 p.

Thornthwaite, C. W.

  1956. Modification of rural microclimates. In "Man's Role in Changing the Face of the Earth," W. L. Thomas, Jr., ed., Part II, p. 567-583. Univ. of Chicago Press, Chicago.

Todd, W. E. C., and M. A. Carriker, Jr.

  1922. The birds of the Santa Marta region of Colombia: a study in altitudinal distribution. Ann. Carnegie Museum 14:1-611.

United Nations Conservation Foundation and the Food and Agricultural Organization
1954. Soil erosion survey of Latin America. J. Soil and Water Conserv. 9:258-280.

United States Department of Interior, United States Board on Geographic Names
1956. Mexico. Gaz. No. 15. Washington. 750 p.

Valenzuela, J.
1939. Informe preliminar de las exploraciones efectuadas en los Tuxtlas, Veracruz. 27th Intern. Congr. of Americanists 2:113-130.

————.

1945. La segunda temporada de exploraciones en la región de los Tuxtlas, estado de Veracruz. An. Inst. Nac. de Antropol. y Hist. 1:81-94, 31 figs

Villaseñor y Sánchez, J. A. de
1746. Teatro Americano, descripcion general de los reynos y provincias de la Nueva Espana y sus jurisdicciones. 2 vols. México, D.F.

Vivó, J. A., and J. C. Gómez
1946. Climatología de México. Inst. Panam. de Geogr. e Hist., Dir. de Geogr., Meteorol., e Hidrol. Publ. 19. 73 p.

Vogt, W.
1945. Unsolved problems concerning wildlife in Mexican national parks. Trans. Tenth N. Am. Wildl. Conf., p. 355-358.

Waibel, L.
1933. Die Sierra Madre de Chiapas. Mitteil. Geogr. Gesell. in Hamburg 43:12-162.

———.

1938. Naturgeschichte der Northers. Geogr. Zeitschr. 44:408-427.

Wallén, C. C.
1955. Some characteristics of precipitation in Mexico. Geogr. Ann. 37:51-85.

———.

1956. Fluctuations and variability in Mexican rainfall. In "The Future of Arid Lands," G. F. White, ed., Am. Assn. Adv. Sci. Publ. No. 43:141-155.

Washington, H. S.
1922. The jade of the Tuxtla statuette. Proc. United States Nat. Museum 60:1-12.

Watters, R. W.
1960. The nature of shifting cultivation: a review of recent research. Pacific Viewpoint 1:59-99.

Wetmore, A.
1941. Notes on birds of the Guatemalan highlands. Proc. United States Nat. Museum 89:523-581.

———.

1942. New forms of birds from Mexico and Colombia. Auk 59:265-268.

———.

1943. The birds of southern Veracruz, Mexico. Proc. United States Nat. Museum 93:215-340.

Weyerstall, A.
1932. Some observations on Indian mounds, idols and pottery in the lower Papaloapan basin, state of Veracruz, Mexico. Tulane Univ., Middle Am. Research Ser. 4:23-69.

Whetten, N. L.
    1948. Rural Mexico. Univ. of Chicago Press, Chicago. 671 p.

Zérega, F.
    1870. El volcán de Tuxtla. Bol. Soc. Geogr. Estad. Rep.
    Mexico 2:500-503.

Zinser, J.
    1936. The Mexican wildlife situation. Trans. First N. Am.
    Wildl. Conf., p. 6-11.

————.

    1944. Mexico's conservation program. Trans Ninth N. Am.
    Wildl. Conf., p. 29-33

UNPLUBLISHED REFERENCES

González, E.
    1962. Unpublished data on game birds and mammals. Forestal Oficina, San Andrés Tuxtla, Veracruz.

Loetscher, F. W., Jr.
    1941. Ornithology of the Mexican state of Veracruz with an annotated list of the birds. Unpublished doctoral thesis, Cornell Univ., Ithaca. 989 p.

Phillips, A.R., and R. W. Dickerman
    1962, 1963. Unpublished ornithological data. Mexico, D.F.

Schieferdecker, A. A. G., and J. H. Tschopp
    1922. Geological report on the Veracruz embayment and the Veracruz isthmus region down to the country of Minatitlan. Unpublished Isthmus Geol. Rep. No. 73, 32 p.

Schumacher, P.
    1929. Geological progress report on San Andres Tuxtla region. Unpublished Geol. Rep. No. 197.

Secretaria de Agricultura y Ganadería
    1952. Ley federal de caza. Dirección General de Caza, Diario Oficial, México, D.F. 7 p.

Staehelin, K. D.
    1935. General report on the Veracruz basin. Unpublished Isthmus Geol. Rep. No. 332, 33 p.

Tschopp, J. H.
    1926. Geological report on lowland between Veracruz and San Andres Tuxtla Mountains. Unpublished Geol. Rep. No. 145, 13 p.

————.

1931. The isthmian saline basin. Unpublished Geol. Rep. No. V - 305.

Appendix A

A NATIONAL PARK OR FOREST AND WILDLIFE PRESERVE

The question of establishing a National Park or Forest and Wildlife Preserve in the Sierra de Tuxtla will, in my opinion, soon become academic if present trends in the region continue. Depletion of the native flora and fauna has been considerable and is progressing with little restraint. Leopold (1959:492) cited the northeast slopes of Volcán San Martín Tuxtla as a location that might well be designated a permanent rain forest reserve. Figure 14 shows how this wilderness section is being encroached upon from all sides and is steadily diminishing in extent.

I analyzed the possibilities for creation of a park or reserve in the region, and I believe there are several valid reasons for establishing such an area here. These are: a) the necessity for conserving the water, forest and soil resources of the Sierra de Tuxtla watershed, b) Mexico's need for more parks on a national level where elements of the original flora and fauna are preserved, c) the suitability of parts of the region for a rain forest reserve in which large mammals, some becoming increasingly rare, can persist in the wild and d) the need for natural areas which can provide certain recreational opportunities for the country's growing population. Each of the above reasons possesses

179

national significance. Based on these reasons and on the future benefits to the region and the country, I believe that the creation of a park is the best use to which part of the Sierra can be put. A decision to preserve some irreplaceable natural features of this region rests in the values and intentions of those in a position to exercise control. The comparative scarcity of good agricultural land in Mexico and the fundamental problems of population expansion and land distribution are some important factors strongly influencing those concerned with the future of the region.

The creation of a park in the Sierra de Tuxtla faces serious obstacles that can be overcome only by a concerted effort. If such a park were established in this area under Article 62 of the Forest Law (Secretaria de Agricultura y Ganadería, 1961:17), it would be necessary to delineate its boundaries as soon as possible by posting and a local display of maps showing its extent. Considerable publicity would be required in the region to explain the reasons for the park's creation and the regulations applying to its use. An adequate staff of rangers with sufficient means of transportation for extensive patrolling would be indispensable for park security. Such measures would be required because there are now few controls in the Sierra where forests and wildlife are concerned. Thus, under present conditions, the creation of a park, though the initial and most important action, would be only the first in a series of essential measures necessary for an effective preserve.

Figure 14 shows the sections of the Sierra which I think would be best suited for park purposes. The core areas would be Volcán San Martín Tuxtla and its slopes to the north and east, and the Volcán Santa Marta - Cerro Campanario complex of north trending ridges including a projection encompassing Volcán San Martín Pajapan. I have selected these sections because a) they contain the largest proportion of continuous primary rain forest in the Sierra, b) in them much of the terrain is better suited for forest than any other use, c) they contain the largest populations of most mammal species, particularly the Baird's Tapir, River Otter and Howler Monkey and d) they are for the most part uninhabited by humans. I include parts of the Gulf coast not only to increase habitat variety, but to preserve some stream environments which River Otters inhabit. Although a part of the wild eastern sections of the Bahía Sontecomapan has been included within the park boundary, I exclude the western and southern shores because of the population concentrations there (Colonia La Palma, Sontecomapan and smaller settlements). For the same reason I omit Lago Catemaco. I think it would be possible, and an asset to the entire park situation, to include either a part or all of both the bay and lake within the park boundaries and jurisdiction so that protection could be extended to the abundant and interesting wildlife in these water and marsh habitats; Lago Catemaco, already a tourist attraction and resort area, could be partly utilized as a recreational section, particularly along its western shore. The inclusion of Bahía Sontecomapan in the park would also make a more cohesive unit by connecting the two large core areas on the massifs.

As outlined on Figure 14 the Volcán San Martín Tuxtla section encompasses approximately 350 square kilometers (35,000 hectares), the Volcán Santa Marta - Cerro Campanario about 475 square kilometers (47,500 hectares). Addition of the whole of Bahía Sontecomapan would add ten square kilometers (1000 hectares), and Lago Catemaco, its northern and eastern shores and the ridge to its north, about 100 square kilometers (10,000 hectares). Thus, a total of 935 square kilometers or 93,500 hectares would make this area, if established under national park status, the second largest unit, and one of the finest, in the Mexican National Park system (Leopold, 1959:91). The possibility can also be considered of setting aside part of the region as strictly a national wildlife refuge under Article 9 of the Federal Game Law (Secretaria de Agricultura y Ganadería, 1952:2), which specifically mentions establishment of refuges for the protection of rare animal species.

The subject of land acquisition and human population within the proposed park boundaries is an important consideration. Articles 45 and 66 of the Forest Law permit the federal government to acquire any land for park purposes. In this event it would be necessary to analyze carefully not only present population distribution, but future settlement possibilities. From population figures and my own observations the number of people at present living within the proposed boundaries of the Volcán San Martín Tuxtla section is slightly less than 1000. The largest populated places (Figure 12) are located on or near the Gulf; these are near Roca Partida, Arroyo del Oro, and at El Real and Montepío, the four comprising 65 per cent of the total population within this section. The population within the other section, excluding the

181

Lago Catemaco addition, is also approximately 1000, with settlements near Punta Zapotitlan, Río Mescalapan and Bahía Sontecomapan comprising about 60 per cent of the total. From these figures it is apparent that a minute proportion of the Sierra's population lives within the boundaries I have outlined. Many of these people have been settled in the area for a comparatively short time. There is probably suitable land outside the proposed park boundary, but still within the region, to which they could move, thus involving a short travel distance. No doubt some could render good service for park patrol, as they are well acquainted with the local surroundings.

This portion of the Sierra is well suited for the establishment of a National Park or National Wildlife Preserve of major proportions. The region possesses natural attributes that would make such a park or refuge an international as well as national scientific and popular attraction. The present use of the region for tourism, recreation and scientific studies would increase with the establishment of a park and contribute toward improving the living standards of the region's people. This would be a significant asset in view of the present economic outlook for the Sierra, one in which industrialization does not figure. The values that would be realized in the preservation of this magnificent natural area and its rich flora and fauna are of great importance to the Sierra, its people and to Mexico. The effort required to accomplish this would be great, but the benefits to be attained make it a worthwhile endeavor.

Appendix B

SOME PLANT SPECIES OF THE SIERRA DE TUXTLA

PLANTS IDENTIFIED FROM THE SIERRA DE TUXTLA

Pterobryaceae (moss family)
    Pterobryum densum (Schwaegr.) Hornsch.

Meteoriaceae (moss family)
    Pilotrichella flexilis (Hedw.) Jaeg.

Polypodiaceae (fern family)
    Pteridium aquilinum var. caudatum (L.) Sadele

Aspidieae (fern family)
    Didymochlaena truncatula (Swartz) J. Smith

Cyatheaceae (tree-fern family)
    Cyathea sp.

Alsophila schiedeana Presl.

Cycadaceae (cycad family)
    Ceratozamia mexicana Brogn.
    Zamia loddigesii var. angustifolia (Regel) Schuster

Podocarpaceae (podocarpus family)
    Podocarpus oleifolius D. Don

Pinaceae (pine family)
    Pinus oocarpa Schiede

Phoenicaceae (palm family)
    Sabal sp.
    Chamaedorea tepejilote Liebm.
    Orbignya sp.
    Astrocaryum mexicanum Liebm.

Musaceae (banana family)
    Heliconia latispatha Benth.

Piperaceae (pepper family)
    Piper auritum H. B. K.
    Piper sp.

Juglandaceae (walnut family)
    Engelhardtia guatemalensis Standl.

Fagaceae (beech family)
    Quercus peduncularis Née, vel aff.
    Quercus oleoides S. & C.
    Quercus ghiesbreghtii Mart. & Gal. vel aff.
    Quercus skinneri Benth.

Ulmaceae (elm family)
    Mirandaceltis monoica (Hemsl.) Sharp

Moraceae (mulberry family)

Ficus glaucescens (Liebm.) Miq.
Ficus cotinifolia
Pseudolmedia oxyphyllaria Donn. Sm.
Cecropia sp.

Urticaceae (nettle family)
Urera elata (Sw.) Griseb.
Myriocarpa longipes Liebm.

Magnoliaceae (magnolia family)
Talauma mexicana (DC.) G. Don

Annonaceae (custard-apple family)
Annona sp.

Myristicaceae (nutmeg family)
Virola guatemalensis (Hemsl.) Warb.

Lauraceae (laurel family)
Phoebe mexicana Meissner

Hamamelidaceae (witch-hazel family)
Liquidambar styraciflua L.

Mimosaceae (mimosa family)
Calliandra grandiflora (L'Her.) Benth.
Pithecollobium arboreum (L.) Urb.

Fabaceae (bean family)
Dussia mexicana (Standl.) Harms
Gliricidea sepium (Jacq.) Steud.
Erythrina americana Mill.

Burseraceae (torchwood family)
Bursera simaruba (L.) Sarg. sens. lat.

Malpighiaceae (malpighia family)
Byrsonima crassifolia (L.) DC.

185

Anacardiaceae (cashew family)
  Spondias mombin L.

Aquifoliaceae (holly family)
  Ilex nitida (Vahl.) Maxim.
  Ilex discolor Hemsl.

Icacinaceae (icacina family)
  Calatola sp.

Elaeocarpaceae (elaeocarpus family)
  Sloanea sp.

Tiliaceae (linden family)
  Apeiba tibourbou Aubl.

Malvaceae (mallow family)
  Hibiscus tiliaceus L.

Bombacaceae (cotton-tree family)
  Bernoullia flammea Oliver
  Pachira aquatica Aubl.

Dilleniaceae (dillenia family)
  Saurauia sp.
  Curatella americana L.

Guttiferae (clusia family)
  Clusia salvinii Donn. Sm.

Flacourtiaceae (flacourtia family)
  Pleuranthodendron mexicana (A. Gray) L. Wms.
  Xylosma sp.

Rhizophoraceae (mangrove family)
  Rhizophora mangle L.

186

Melastomaceae (meadow-beauty family)
Conostegia xclapensis (Bonpl.) DC.

Araliaceae (ginseng family)
Oreopanax capitatum (Jacq.) D. & P.
Oreopanax xalapense (H. B. K.) D. & P.
Dendropanax arboreum (L.) D. & P.

Clethraceae (white-alder family)
Clethra suaveolens Turcz.
Clethra macrophylla M. & G.

Apocynaceae (dogbane family)
Stemmadenia galeottiana (A. Rich.) Miers.

Boraginaceae (borage family)
Cordia spinescens L.

Acanthaceae (acanthus family)
Aphelandra aurantiaca (Schiedw.) Lindl.
Odontonema callistachyum (S. & C.) Kuntze.

Rubiaceae (madder family)
Hamelia longipes Standl.
Hamelia patens Jacq.
Hoffmannia lenticillata Hemsl.
Cephaelis elata Sw.

Caprifoliaceae (honeysuckle family)
Viburnum acutifolium Benth.

Compositae (composite family)
Melampodium divaricatum (Rich.) DC.
Senecio sp.
Polymnia maculata Cav.

LOCAL NAMES OF PLANTS IN THE SIERRA DE TUXTLA

Achiote. Bixa oxellana.
Aguacate. Persea americana.
Algodocillo.
Amarillo.
Amate. Ficus sp.
Amate cajonero. Ficus sp.
Anacahuite.
Apipi.
Arrayan. Laurus vulgaris.
Bejuco.
Cahani.
Cañafistula. Cassia grandis.
Caoba. Swietenia sp.
Caobilla. Croton lucidum. = [*Croton glabellus.*]*
*Capulin. Muntingia calabura.**
Cedro. Cedrela mexicana.
Cedro nogal.
Ceiba. Ceiba pentandra.
*Cipres de bálsamo. Toluifera pereirae.**
*Ciruela amarilla. Spondias lutea.**
*Ciruela colorada. Spondias mombin.**
Cocuite. Ichthyomethia communis.
Copite.
Cuero.
Chagani.
Chico zapote. Achras zapota.
Chicozapote. Dyospyros obtusifolia. = [*Diospyros ebenaster.*]
Dagame. Calycophyllum sp.*
Ebano.
Encino amarillo. Quercus sp.
Encino blanco. Quercus sp.
Encino prieto. Quercus sp.
Escobilla.
Fresno. Rhus sp.

Gaga.
*Gateado. Swietenia humilis.**
Guanábana. Annona sp.
Guasímo. Guazuma ulmifolia.
Guayaba. Psidium sp.
Hule. Castilla elastica.
Ixtle. Bromelia silvestris.
Jaboncillo. Dussia mexicana.
Jicama.
Jicarillo.
Jinicuil.
*Jobo o jonote. Spondias mombin.**
Jonote. Heliocarpus tomentosus.
Lagarto.
Laurel.
Laurellio.
Limoncillo.
Liquidambar. Liquidambar styraciflua.
Llolo. Talauma mexicana.
Macayo. Andira sp.
Maguey. Agave americana.
Majagua. Hibiscus sp.
*Mangles blanco. Laguncularia racemosa.**
*Mangles colorado. Rhizophora mangle.**
*Mangles negro. Conocarpus erecta.**
Moral. Chlorophora tinctoria.
Mulato. Elaphrium simaruba. = [*Bursera simaruba.*]
Nanche. Byrsonima crassifolia.
Naranjo del monte.
Nazareno. Brosimum sp.*
Nogal.
Ocote. Pinus sp.
Ocote amarillo. Pinus variabilis. = [*Pinus oocarpa.*]*
Ojochin. Brosimum sp.*
Palma coyol. Orbignya sp.
Palma de coco. Cocos nucifera.
Palma de chocho. Astrocaryum mexicanum.
Palma de tepejilote. Chamaedorea tepejilote.

Palma real. Sabal mexicana.*
Palma redonda. Sabal mexicana.*
Palo blanco.
Palo de agua. Astianthus viminalis.
Palo de bejuco.
Palo de Campeche. Haematoxylum sp.*
Palo de cuchara.
Palo muladoro.
Palo mulato. Bursera simaruba.*
Palo verde. Ilex sp.
Papachota. Arundinella hispida.
Paro amarillo. Morus tinctorea.
Pimienta.
Plantanillo.
Rabo.
Roble blanco. Tabebuia sp.
Soncuavite. Pithecollobium arboreum.
Suchil. Plumeria acutifolia.*
Tabies.
Tamalcuavite.
Tamarindo. Tamerindus occidentalis.
Tepejilote. Chamaedorea tepejilote.
Teposi.
Tepozontle.
Uvero. Coccoloba sp.
Xoxogo.
Zapote agrio.
Zapote mamey. Mammea americana.
Zarza. Buettneria sp.*

* probable scientific name added.

Sources:
Andrle (field work, 1962)
Boettiger de Alvarez (1961a, 1961b)
Covarrubias (1946)
Escobar (1794)
Foster (1942)

Gomez Mayorga (1961)
González (1962)
Jaussoro (1961)
Medel y Alvarado (1963)
"Santiago Tuxtla, Rincón de Maravilla" (1961)
Standley (1920-1926)

Appendix C

NONTRANSIENT BIRD SPECIES OF THE SIERRA DE TUXTLA

RELATIVE ABUNDANCE AND FREQUENCY OF OCCURRENCE CATEGORIES

Based on sight and/or sound record in preferred habitat of species.

Abundant - recorded in considerable numbers every day (over 150).

Very common - recorded in moderate numbers every day (75-150).

Common - recorded in limited numbers every day (15-75).

Fairly common - recorded in low numbers every day (5-15).

Rather uncommon - recorded frequently but normally in very low numbers and not every day (3-8).

Uncommon - recorded sparingly even in the most suitable places and more likely to be not seen than observed on a given day (1-5).

Very uncommon - recorded infrequently and usually not observed for one to three weeks at a time (normally 1-5 in one to three months).

Rare - recorded very infrequently or not at all for varying, sometimes prolonged, periods (normally not more than 1-10 per year).

Very rare - recorded from one to three or four times in the region and likely to occur again.

PARTIALLY ANNOTATED LIST BY HABITAT

Included are relative abundance and altitudinal range above sea level. A hyphen before or after a single elevation or after the upper limit of a species' range indicates that lower or upper limits are not known and that a species probably occurs beyond the elevations given. Eisenmann (1955) has been followed for most vernacular and scientific names. First records for the study area that have not previously been published are indicated.

Species characteristic of tropical rain forest and cloud forest.

Tinamus major. Great Tinamou. Uncommon; 0-850 meters.

Crypturellus soui. Little Tinamou. Uncommon; 0-800 meters.

Crypturellus boucardi. Slaty-breasted Tinamou. Fairly common; 250-900 meters.

194

Chondrohierax uncinatus. Hook-billed Kite. Very rare; 300 meters; first record. The only record of this species within the Sierra is a male collected by A. Ramírez V. near Dos Amates on Dec. 5, 1961 (rep. *in litt.* from A. R. Phillips).

Harpagus bidentatus. Double-toothed Kite. Very rare; 350 meters; first record. H. Axtell and I closely observed an individual near Dos Amates on March 17, 1960.

Accipiter bicolor. Bicolored Hawk. Very rare; ca. 400 meters. Edwards and Tashian (1959:328) recorded this species near Coyame in the summer of 1954, and Ramírez collected an immature near Dos Amates in late 1962 (rep. *in litt.* from A. R. Phillips).

Leucopternis albicollis. White Hawk. Rather uncommon; 0-900 meters. Also observed in drier semideciduous forest.

Spizaetus tyrannus. Black Hawk-Eagle. Very rare; 0-850 meters; first record. Also observed in drier semideciduous forest.

Micrastur semitorquatus. Collared Forest-Falcon. Uncommon; 0-1300 meters; first record.

Micrastur ruficollis. Barred Forest-Falcon. Rather uncommon; 500-1500 meters; first record.

Crax rubra. Great Curassow. Rather uncommon; 0-900 meters.

Penelope purpurascens. Crested Guan. Fairly common; 0-1400 meters.

Odontophorus guttatus. Spotted Wood-Quail. Fairly common; 300-1300 meters.

Columba speciosa. Scaled Pigeon. Rather uncommon; 350-800 meters.

Columba nigrirostris. Short-billed Pigeon. Fairly common; 0-750 meters.

Leptotila verreauxi. White-tipped Dove. Common; 0-1100 meters. Also observed in drier semideciduous forest.

Leptotila plumbeiceps. Gray-headed Dove. Fairly common; 0-750 meters.

Geotrygon lawrencii. Purplish-backed Quail-Dove. Fairly common; 350-1400 meters.

Geotrygon montana. Ruddy Quail-Dove. Rather uncommon; 0-900 meters.

Bolborhynchus lineola. Barred Parakeet. Vary rare; 700 meters; first record. Five were reported in the upper edge of the pine forest above Ocotal Chico on Dec. 7, 1962 by Phillips and one on Dec. 11 by an assistant.

Pionopsitta haematotis. Brown-hooded Parrot. Very rare; ca. 400 and 550 meters. This parrot has been recorded only by Edwards and Tashian (1959:328) near Coyame and Davis (1952:315) south of Lago Catemaco.

Amazona autumnalis. Red-lored Parrot. Common (local); 0-700 meters. Also observed in drier semideciduous forest.

Otus guatemalae. Vermiculated Screech Owl. Uncommon; 0-1100 meters. Also observed in drier semideciduous forest.

Pulsatrix perspicillata. Spectacled Owl. Very rare; 0 meters. The sole report is from Sontecomapan by Sclater (1857:227) (*Ciccaba torquata*).

Ciccaba virgata. Mottled Owl. Fairly common; 0-1600 meters. Also observed in drier semideciduous forest.

Phaethornis superciliosus. Long-tailed Hermit. Fairly common; 0-850 meters.

196

Phaethornis longuemareus. Little Hermit. Rather uncommon; 0-550 meters.

Campylopterus curvipennis. Wedge-tailed Sabrewing. Fairly common; 0-1200 meters.

Campylopterus hemileucurus. Violet Sabrewing. Rather uncommon; 0-1500 meters.

Colibri thalassinus. Green Violet-ear. Uncommon (local); 650-800 meters; first record. I found this species only in the humid forest on the south slope of Volcán San Martín Tuxtla from 650-800 meters, where large oaks are present. I collected a male and female on June 4 and 6, 1962.

Amazilia candida. White-bellied Emerald. Fairly common; 0-1100 meters.

Lampornis amethystinus. Amethyst-throated Hummingbird. Very rare; 1280 meters; first record. I secured a female, the only record, on Aug. 26, 1962, in the primary forest on Volcán San Martín Tuxtla.

Trogon massena. Slaty-tailed Trogon. Uncommon; 0-500 meters.

Trogon collaris. Bar-tailed Trogon. Fairly common; 0-1650 meters.

Trogon violaceus. Violaceus Trogon. Fairly common; 0-1000 meters. Also observed in drier semideciduous forest.

Hylomanes momotula. Tody Motmot. Rare; -750 meters.

Momotus momota. Blue-crowned Motmot. Fairly common; 0-1200 meters. Also observed in drier semideciduous forest.

Galbula ruficauda. Rufous-tailed Jacamar. Uncommon; 0- meters; first record. I observed this species only in the vicinity of the Río

197

Carizal on the Gulf side of the range. It is possibly resident at other points near the coast.

Aulacorhynchus prasinus. Emerald Toucanet. Rather uncommon; 450-1400 meters.

Pteroglossus torquatus. Collared Aracari. Rather uncommon; 0-600 meters. Also observed in drier semideciduous forest.

Piculus rubiginosus. Golden-olive Woodpecker. Rather uncommon; -1600 meters.

Celeus castaneus. Chestnut-colored Woodpecker. Very uncommon; 0-600 meters.

Centurus pucherani. Black-cheeked Woodpecker. Very uncommon; 0-800 meters.

Phloeceastes guatemalensis. Pale-billed Woodpecker. Uncommon; 0-950 meters. Also observed in drier semideciduous forest.

Dendrocincla anabatina. Tawny-winged Woodcreeper. Very uncommon; 0-800 meters.

Sittasomus griseicapillus. Olivaceous Woodcreeper. Fairly common; -900 meters.

Dendrocolaptes certhia. Barred Woodcreeper. Uncommon; -800 meters.

Xiphorhynchus flavigaster. Ivory-billed Woodcreeper. Fairly common; 0-1250 meters. Also observed in drier semideciduous forest.

Lepidocolaptes souleyetii. Streak-headed Woodcreeper. Rather uncommon; -800- meters. Also observed in drier semideciduous forest.

Lepidocolaptes affinis. Spot-crowned Woodcreeper. Uncommon; 900-1600 meters.

Anabacerthia variegaticeps. Scaly-throated Foliage-gleaner. Uncommon; 700-1200 meters.

Automolus ochrolaemus. Buff-throated Foliage-gleaner. Fairly common; 0-1400 meters.

Xenops minutus. Plain Xenops. Uncommon; 0-850 meters.

Formicarius analis. Black-faced Antthrush. Fairly common; 0-950 meters.

Grallaria guatimalensis. Scaled Antpitta. Rare; 0-800- meters.

Pipra mentalis. Red-capped Manakin. Very uncommon; 0-850 meters.

Cotinga amabilis. Lovely Cotinga. Very rare; 300 meters; first record. Ramírez collected two specimens, an adult and an immature, on Jan. 7 and 8, 1961, respectively, near Dos Amates. (rep. *in litt.* from A. R. Phillips).

Attila spadiceus. Bright-rumped Attila. Rather uncommon; 0-950 meters. Also observed in drier semideciduous forest.

Empidonax flavescens. Yellowish Flycatcher. Rather uncommon; 900-1600 meters. I observed this flycatcher only on Volcán San Martín Tuxtla.

Myiobius sulphureipygius. Sulphur-rumped Flycatcher. Very uncommon; 0-600 meters.

Onychorhynchus mexicanus. Northern Royal Flycatcher. Rare; 0-650 meters; first record. Ramírez collected several specimens, the first on Dec. 19, 1960. near Dos Amates. (rep. *in litt.* from A. R. Phillips).

Platyrinchus mystaceus. White-throated Spadebill. Rather uncommon; 0-825 meters.

Tolmomyias sulphurescens. Yellow-olive Flycatcher. Uncommon; 0-750 meters. Also observed in drier semideciduous forest.

Rhynchocyclus brevirostris. Eye-ringed Flatbill. Rather uncommon; 500-1200 meters.

Oncostoma cinereigulare. Northern Bentbill. Very uncommon; 0-850 meters.

Ornithion semiflavum. Yellow-bellied Tyrannulet. Rare; 0-800- meters; first record.

Leptopogon amaurocephalus. Sepia-capped Flycatcher. Very rare; 300 meters; first record. The only Sierra record is a female secured by Ramírez at Dos Amates on Dec. 4, 1961.

Pipromorpha oleaginea. Ochre-bellied Flycatcher. Rather uncommon; -950 meters.

Cyanocorax yncas. Green Jay. Fairly common; 250-1300 meters. Also observed in drier semideciduous forest.

Henicorhina leucosticta. White-breasted Wood-Wren. Common; 0-1500- meters.

Turdus assimilis. White-throated Robin. Common; 300-1300 meters.

Turdus infuscatus. Black Robin. Rare; 1300-1600 meters; first record. I secured a male at 1600 meters elevation in the elfin forest near the crater of Volcán San Martín Tuxtla. One or two others were observed at about 1300 meters on this volcano.

Myadestes unicolor. Slate-colored Solitaire. Fairly common; 450-1650 meters.

Catharus mexicanus. Black-headed Nightingale-Thrush. Fairly common; 650-1400 meters.

Catharus aurantiirostris. Orange-billed Nightingale-Thrush. Very rare; 600 meters; first record. I observed this species only once, an individual on March 22, 1960, above Dos Amates.

Smaragdolanius pulchellus. Green Shrike-Vireo. Rare; 0-800- meters; first record.

Vireo olivaceus. Red-eyed (Yellow-green) Vireo. Common; 0-1250 meters. Also observed in drier semideciduous forest.

Hylophilus ochraceiceps. Tawny-crowned Greenlet. Rather uncommon; 0-900 meters.

Hylophilus decurtatus. Gray-headed Greenlet. Fairly common; 0-850 meters.

Cyanerpes cyaneus Red-legged Honeycreeper. Common; 0-950 meters. Also observed in drier semideciduous forest.

Parula pitiayumi. Tropical Parula Warbler. Fairly common; 150-950 meters. This warbler is a resident both in the humid and drier forests of the Sierra. The Distributional Check-List of the Birds of Mexico (1957:243) records it as only casual in northeast Oaxaca and central Veracruz.

Myioborus miniatus. Slate-throated Redstart. Fairly common; 600-1250 meters.

Basileuterus culicivorus. Golden-crowned Warbler. Common; 0-1000 meters.

Basileuterus belli. Golden-browed Warbler. Uncommon; 750-1600 meters.

Amblycercus holosericeus. Yellow-billed Cacique. Fairly common; 0-600 meters. Also in drier semideciduous forest.

Chlorophonia occipitalis. Blue-crowned Chlorophonia. Very rare; 300,

600, 750 meters. The only records are an unlabeled specimen said by Ramírez to have been taken above Dos Amates, a pair collected on Jan. 18, 1948, near San Andrés Tuxtla (Lowery and Dalquest, 1951:635), a male which I collected on June 5, 1962, at about 750 meters on the south slope of Volcán San Martín Tuxtla, and a specimen taken by Ramírez near Dos Amates on Jan. 2, 1964 (rep. *in litt.* from A. R. Phillips).

Piranga leucoptera. White-winged Tanager. Uncommon; 500-1000-meters.

Habia rubica. Red-crowned Ant-Tanager. Rather uncommon; 0-1100 meters. Also observed in drier semideciduous forest.

Habia gutturalis. Red-throated Ant-Tanager. Fairly common; 0-850 meters. Also observed in drier semideciduous forest.

Lanio aurantius. Black-throated Shrike-Tanager. Very uncommon; 0-650 meters.

Eucometis penicillata. Gray-headed Tanager. Uncommon; -650 meters.

Chlorospingus ophthalmicus. Common Bush-Tanager. Fairly common; 500-1600 meters.

Caryothraustes poliogaster. Black-faced Grosbeak. Common; 0-850 meters.

Cyanocompsa cyanoides. Blue-black Grosbeak. Rather uncommon; 0-850 meters.

Atlapetes brunnei-nucha. Chestnut-capped Brush-Finch. Fairly common; 300-1650 meters.

Species which occur principally in forest edge, thickets, bush and tree rows and fields in humid areas.

Buteo magnirostris. Roadside Hawk. Fairly common; 0-850 meters. Also in drier areas.

Buteo nitidus. Grey Hawk. Rather uncommon; 0-950 meters. Also in drier areas.

Herpetotheres cachinnans. Laughing Falcon. Very uncommon; 0-650 meters. Also in drier areas.

Falco albigularis. Bat Falcon. Very uncommon; 0-800 meters. Also in drier areas.

Ortalis vetula. Plain Chachalaca. Common; 0-1000 meters. Also in drier areas.

Columba flavirostris. Red-billed Pigeon. Common; 0-1660 meters. Also in drier areas.

Columbigallina minuta. Plain-breasted Ground-Dove. Very rare; ca. 500 meters. Edwards and Tashian (1959:330) reported the only Sierra record near Coyame in 1954.

Columbigallina talpacoti. Ruddy Ground-Dove. Common; 0-800 meters. Also in drier areas.

Claravis pretiosa. Blue Ground-Dove. Uncommon; 0-950 meters.

Aratinga astec. Olive-throated Parakeet. Common; 0-750 meters. Also in drier areas.

Coccyeus minor. Mangrove Cuckoo. Very rare; 625 meters; first record. Also in drier areas(?). I observed one of these cuckoos on June 6, 1962, in an area of cornfields and thickets at about 625 meters elevation several kilometers south of Volcán San Martín Tuxtla.

Piaya cayana. Squirrel Cuckoo. Rather uncommon; 0-750 meters. Also in drier areas.

Crotophaga sulcirostris. Groove-billed Ani. Common; 0-650 meters. Also in drier areas.

Tapera naevia. Striped Cuckoo. Very uncommon; 0-350 meters. Also in drier areas.

Glaucidium brasilianum. Ferruginous Pygmy Owl. Rather uncommon; 0-750 meters. Also in drier areas.

Nyctibius griseus. Common Potoo. Rare; 0-800 meters. Also in drier areas.

Nyctidromus albicollis. Pauraque. Common; 0-950 meters. Also in drier areas.

Anthracothorax prevostii. Green-breasted Mango. Uncommon; 0-650 meters. Also in drier areas.

Paphosia helenae. Black-crested Coquette. Very uncommon; 600- meters; first record. The only individuals I observed were in pine and in humid tropical forest in the Ocotal area. Two males were secured on May 16 and Oct. 23, 1962. Phillips collected three females from the same vicinity in Dec. 1962.

Chlorostilbon canivetii. Fork-tailed Emerald. Rather uncommon; 0-800 meters. Also in drier areas.

Amazilia tzacatl. Rufous-tailed Hummingbird. Fairly common; 0-800 meters. Also in drier areas.

Ramphastos sulfuratus. Keel-billed Toucan. Fairly common; 0-1250 meters. Also in drier areas.

Centurus aurifrons. Golden-fronted Woodpecker. Common; 0-1000-meters. Also in drier areas.

Veniliornis fumigatus. Smoky-brown Woodpecker. Very uncommon; 0-800 meters.

Synallaxis erythrothorax. Rufous-breasted Spinetail. Fairly common; 0-750 meters. Also in drier areas.

Taraba major. Great Antshrike. Very rare; 0- meters. Wetmore (1943:282) reported this species at the base of the Sierra. The only records within my study area are those by Sclater (1857:203) from Sontecomapan and Ramírez, a male and female collected from the vicinity of Dos Amates (300 meters) in the latter part of Dec. 1962 (rep. *in litt.* from A. R. Phillips).

Thamnophilus doliatus. Barred Antshrike. Fairly common; 0-850 meters. Also in drier areas.

Pachyramphus major. Gray-collared Becard. Rare; 0-1250 meters. Also in drier areas.

Platypsaris aglaiae. Rose-throated Becard. Rather uncommon; 0-750 meters. Also in drier areas.

Tityra semifasciata. Masked Tityra. Fairly common; 0-1250 meters. Also in drier areas.

Tityra inquisitor. Black-crowned Tityra. Very uncommon. 0-300 meters; first record. Also in drier areas.

Tyrannus melancholicus. Tropical Kingbird. Common; 0-950 meters. Also in drier areas.

Legatus leucophaius. Piratic Flycatcher. Rather uncommon; 0-900 meters.

Myiodynastes luteiventris. Sulphur-bellied Flycatcher Fairly common;

205

0-950 meters. Also in drier areas.

Myiodynastes maculatus. Streaked Flycatcher. Rather uncommon; 0-750 meters.

Megarynchus pitangua. Boat-billed Flycatcher. Fairly common; 0-850 meters. Also in drier areas.

Myiozetetes similis. Vermilion-crowned Flycatcher. Common; 0-850 meters. Also in drier areas.

Pitangus sulphuratus. Great Kiskadee. Common; 0-850 meters. Also in drier areas.

Myiarchus tyrannulus. Brown-crested Flycatcher. Fairly common; 0-750 meters. Also in drier areas.

Myiarchus tuberculifer. Dusky-capped Flycatcher. Fairly common; 0-1300 meters. Also in drier areas.

Contopus cinereus. Tropical Pewee. Uncommon; -150-700 meters. Also in drier areas.

Elainea flavogaster. Yellow-bellied Elainia. Uncommon; 0-850 meters. Also in drier areas.

Myiopagis viridicata. Greenish Elainia. Very uncommon; 0-600 meters; first record. Also in drier areas.

Psilorhinus morio. Brown Jay. Common; 0-800 meters. Also in drier areas.

Campylorhynchus zonatus. Band-backed Wren. Fairly common; 0-650 meters. Also in drier areas.

Thryothorus maculipectus. Spot-breasted Wren. Fairly common; 0-1250 meters. Also in drier areas.

Troglodytes musculus. Southern House Wren. Uncommon; -100-850 meters.

Turdus grayi. Clay-colored Robin. Common; 0-950 meters. Also in drier areas.

Ramphocaenus rufiventris. Long-billed Gnatwren. Very uncommon; 0-850 meters. Also in drier areas.

Cyclarhis gujanensis. Rufous-browed Peppershrike. Uncommon; 0-600 meters. Also in drier areas.

Coereba flaveola. Bananaquit. Rather uncommon; 0-1100 meters.

Chamaethlypis poliocephala. Gray-crowned Yellowthroat. Fairly common; 0-750- meters. Also in drier areas.

Basileuterus rufifrons. Rufous-capped Warbler. Fairly common; 0-950- meters.

Zarhynchus wagleri. Chestnut-headed Oropendola. Very rare; -350 meters. The only records are Sclater (1857:228) from San Andrés Tuxtla and a male and female collected at Dos Amates by Ramírez on Nov. 23 and 14, respectively, in 1961.

Gymnostinops montezuma. Montezuma Oropendola. Fairly common; 0-650 meters.

Psomocolax oryzivorus. Giant Cowbird. Very uncommon; 0-600 meters.

Tangavius aeneus. Red-eyed Cowbird. Fairly common; 0-700 meters. Also in drier areas.

Cassidix mexicanus. Boat-tailed Grackle. Very common; 0-800 meters. Also in drier areas.

Dives dives. Melodius Blackbird. Common; 0-700 meters. Also in drier areas.

Icterus prosthemelas. Black-cowled Oriole. Rather uncommon; 0-750 meters. Also in drier areas.

Icterus wagleri. Wagler's Oriole. Very rare; 500 meters. Davis (1952:315) reports this species south of Lago Catemaco.

Icterus cucullatus. Hooded Oriole. Rare; 0- meters; first record. Dickerman reported four in forest edge north of Sontecomapan on August 7, 1963.

Icterus mesomelas. Yellow-tailed Oriole. Uncommon; 0-650 meters.

Tanagra musica. Blue-hooded Euphonia. Very rare; 500-850 meters; first record. On March 22, 1960, I noted a pair above Dos Amates and observed another individual on Oct. 30, 1962, above Ocotal Chico. Phillips reported about eight and collected a male above Ocotal Chico on Dec. 8, 1962, and received a specimen taken by Ramírez near Dos Amates on Nov. 6, 1963.

Tanagra affinis. Scrub Euphonia. Rather uncommon; -850 meters. Also in drier areas.

Tanagra lauta. Yellow-throated Euphonia. Fairly common; 0-850 meters. Also in drier areas.

Tanagra gouldi. Olive-backed Euphonia. Very uncommon; -700 meters; first record.

Thraupis virens. Blue-gray Tanager. Fairly common; 0-750 meters. Also in drier areas.

Thraupis abbas. Yellow-winged Tanager. Common; 0-850 meters. Also in drier areas.

Phlogothraupis sanguinolenta. Crimson-collared Tanager. Rather uncommon; 0-650 meters.

Saltator atriceps. Black-headed Saltator. Common; 0-900 meters. Also in drier areas.

Saltator maximus. Buff-throated Saltator. Very uncommon; 0-650 meters.

Saltator coerulescens. Grayish Saltator. Fairly common; 0-750 meters. Also in drier areas.

Richmondena cardinalis. Common Cardinal. Rather uncommon; 0-850 meters. Also in drier areas.

Cyanocompsa parellina. Blue Bunting. Very uncommon; 0-650 meters. Also in drier areas.

Tiaris olivacea. Yellow-faced Grassquit. Fairly common; 0-750- meters. Also in drier areas.

Sporophila torqueola. White-collared Seedeater. Common; 0-750- meters. Also in drier areas.

Volatinia jacarina. Blue-black Grassquit. Common; 0-650- meters. Also in drier areas.

Spinus psaltria. Dark-backed Goldfinch. Uncommon; 200-900 meters; first record. Also in drier areas.

Arremonops rufivirgatus. Olive Sparrow. Fairly common; 0-850 meters. Also in drier areas.

Aimophila rufescens. Rusty Sparrow. Fairly common; 0-950- meters. Also in drier areas.

Species characteristic of semideciduous forest, forest edge, thickets and bush and tree rows in drier areas.

Crypturellus cinnamomeus. Rufescent Tinamou. Rather uncommon; 0-800 meters.

Colinus virginianus. Common Bobwhite. Fairly common; 0-800 meters. First record.

Zenaida asiatica. White-winged Dove. Rare; 0-600 meters. Also in more humid areas.

Scardafella inca. Inca Dove. Uncommon; 0-450 meters. Also in more humid areas.

Columbigallina passerina. Common Ground-Dove. Very rare; 350 meters. The only Sierra record is that of Sclater (1857:205) from San Andrés Tuxtla.

Amazona ochrocephala. Yellow-headed Parrot. Very uncommon; ca. 250 meters; first record. Phillips reported at least three on Dec. 16, 1962, in the oak forest area south of Soteapan.

Coccyzus americanus. Yellow-billed Cuckoo. Uncommon; 100-200 meters; first record. I observed from one to three individuals in the oak and semideciduous forests on the southern side of the range on June 16, 17, and July 3, 1962. A female collected on the 17th near Barrosa had the gonad slightly enlarged.

Rhinoptynx clamator. Striped Owl. Very rare; 150-300 meters; first record. Also in more humid areas. The only records are a female which I collected in semideciduous open woodland at Barrosa on June 16, 1962, and a female with two young collected by Ramírez near Dos Amates in early 1963 (rep. *in litt.* from A. R. Phillips).

Amazilia yucatanensis. Fawn-breasted Hummingbird. Uncommon; 0-400 meters. Also in more humid areas.

Heliomaster longirostris. Long-billed Starthroat. Uncommon; 0-800 meters; first record. Also in more humid areas

Trogon citreolus. Citreoline Trogon. Fairly common; 0-700 meters. Also in more humid areas.

Dryocopus lineatus. Lineated Woodpecker. Uncommon; 0-700 meters. Also in more humid areas.

Dendrocopos scalaris. Ladder-backed Woodpecker. Uncommon; 0-700-meters. Also in more humid areas.

Pyrocephalus rubinus. Vermilion Flycatcher. Very uncommon; 0-400 meters. Also in more humid areas.

Muscivora tyrannus. Fork-tailed Flycatcher. Very uncommon; 0-700 meters. Also in more humid areas.

Camptostoma imberbe. Beardless Flycatcher. Uncommon; 0-400 meters; first record. Also in more humid areas.

Uropsila leucogastra. White-bellied Wren. Rather uncommon; 0-600 meters. Also in more humid areas.

Mimus polyglottos. Common Mockingbird. Very rare; 400 meters; first record. I observed what appeared to be a typical individual of *M. polyglottos* on the northwest shore of Lago Catemaco April 2, 1960.

Polioptila caerulea. Blue-gray Gnatcatcher. Rather uncommon; 0-200 meters; first record.

Icterus gularis. Black-throated Oriole. Uncommon; 0-700 meters. Also in more humid areas.

211

Species characteristic of pine and oak forest.

Amazilia cyanocephala. Red-billed Azure-crown. Fairly common; 500-850- meters; first record. Also occurs in other habitats. This hummingbird inhabits chiefly pine forests and edges of gum-oak forest south of Volcán Santa Marta where I collected a male on May 15, 1962, at 825 meters.

Melanerpes formicivorus. Acorn Woodpecker. Common; 0-850- meters; first record. Also occurs in other habitats.

Parus atricristatus. Black-crested Titmouse. Very uncommon; 150 meters; first record. The only one I observed was in the open oak forests near Guasuntlan on May 14, 1962.

Icterus chrysater. Yellow-backed Oriole. Uncommon; 0-750 meters. Also in other habitats.

Species which range widely in open humid and drier areas.

Sarcoramphus papa. King Vulture. Very uncommon; 0-900 meters.

Coragyps atratus. Black Vulture. Very common; 0-1660 meters.

Cathartes aura. Turkey Vulture. Rather uncommon; 0-1660 meters.

Buteo albonotatus. Zone-tailed Hawk. Very rare; 400 meters; first record. I observed one at close range on Cerro Blanco on Sept. 20, 1962.

Buteo brachyurus. Short-tailed Hawk. Very rare; 0-800 meters; first record. A dark phase individual was noted over the northwest shore of Lago Catemaco on Oct. 11, 1962, and a light phase

212

bird above Ocotal Chico at 800 meters on Oct. 25, 1962. Dickerman reported a dark phase bird and possibly a light phase one in 1963 on Aug. 12 and 16, respectively, north-northeast of Sontecomapan.

Polyborus cheriway. Crested Caracara. Rather uncommon; 0-750 meters.

Falco femoralis. Aplomado Falcon. Very uncommon; 0-600 meters.

Tyto alba. Barn Owl. Very uncommon; 150-600 meters; first record.

Chordeiles acutipennis. Lesser Nighthawk. Rare; 0-500 meters; first record.

Streptoprocne zonaris. White-collared Swift. Fairly common; 0-1000 meters.

Chaetura vauxi. Vaux's Swift. Common; 0-1100 meters.

Aeronautes saxatalis. White-throated Swift. Very rare; 350, 800 meters; first record. I observed this species over the Sierra on only two occasions. Two birds on Oct. 29, 1962, flew past above Ocotal Grande at about 800 meters. Another passed over the northwest shore of Lago Catemaco on Nov. 14, 1962.

Progne chalybea. Gray-breasted Martin. Uncommon; 0-800 meters.

Stelgidopteryx ruficollis. Rough-winged Swallow. Fairly common; 0-800 meters.

Sturnella magna. Common Meadowlark. Very rare; -250 meters; first record. The species is fairly common in the lowlands but apparently seldom reaches higher elevations.

213

Species which are usually associated with water bodies in the form of sea, lakes, swamps, streams.

Podiceps dominicus. Least Grebe. Common; 0-400 meters; first record.

Podilymbus podiceps. Pied-billed Grebe. Fairly common; 0-400 meters.

Pelecanus occidentalis. Brown Pelican. Uncommon; 0-350 meters.

Phalacrocorax olivaceus. Olivaceous Cormorant. Abundant (Lago Catemaco); 0-400 meters.

Anhinga anhinga. Anhinga. Rare; 0-400 meters; first record.

Fregata magnificens. Magnificent Frigatebird. Fairly common; 0-400 meters.

Ardea herodias. Great Blue Heron. Very uncommon; 0-650 meters; first record.

Butorides virescens. Green Heron. Fairly common; 0-650 meters.

Florida caerulea. Little Blue Heron. Uncommon (irregular, occasional flocks); 0-650 meters; first record.

Casmerodius albus. Common Egret. Rare; 0-350 meters; first record.

Leucophoyx thula. Snowy Egret. Abundant (Lago Catemaco); 0-350 meters.

Bubulcus ibis. Cattle Egret. Uncommon (irregular, flocks); 0-450 meters; first record.

Hydranassa tricolor. Tricolored Heron. Very uncommon; 0-350 meters; first record.

Nycticorax nycticorax. Black-crowned Night Heron. Fairly common; 0-600 meters.

Nyctanassa violacea. Yellow-crowned Night Heron. Very uncommon; 0-350 meters; first record.

Ixobrychus exilis. Least Bittern. Very rare; 350 meters; first record. I observed three in the marsh vegetation at the Arroyo Agrio at the north side of Lago Catemaco on June 10, 1962. Dickerman observed from one to four birds in the same place in July and August 1963, and collected two on Aug. 9.

Cochlearius cochlearius. Boat-billed Heron. Very rare; 350 meters; first record. The only record within the study area is that reported by Sclater (1857:230) probably at Catemaco (Cateman).

Mycteria americana. Wood Ibis. Very rare; 350 meters; first record. I saw a number in the lowlands but only one in the Sierra, over Lago Catemaco on April 4, 1960.

Eudocimus albus. White Ibis. Very rare; 0- meters. One was seen in Bahía Sontecomapan on May 25, 1962, and 20 flew past the mouth of the Río Carizal on Aug. 15, 1962. Sclater (1857:230) also recorded it from Sontecomapan.

Dendrocygna autumnalis. Black-bellied Tree-Duck. Rare; (Lago Catemaco) 350 meters; first record.

Oxyura jamaicensis. Ruddy Duck. Rare (flocks on Lago Catemaco); 350 meters; first record.

Elanus leucurus. White-tailed Kite. Very uncommon; 0-750 meters.

Rostrhamus sociabilis. Snail Kite. Rather uncommon; 0-350 meters; first record.

Buteogallus anthracinus. Common Black Hawk. Rather uncommon; 0-1100 meters.

Hypomorphnus urubitinga. Great Black Hawk. Uncommon; 0-800 meters, first record.

Pandion haliaetus. Osprey. Very uncommon; 0-250 meters; first record.

Aramus guarauna. Limpkin. Very rare; 0- meters; first record. I observed one fly up the Río Coxcoapan at its mouth in Bahía Sontecomapan on March 29, 1962.

Amaurolimnus concolor. Uniform Crake. Very rare; 0- meters; first record. Ramírez collected an adult male on February 15, 1963 in Bahía Sontecomapan (rep. *in litt.* from R. W. Dickerman).

Aramides cajanea. Gray-necked Wood-Rail. Rare; 0- meters; first record. The only ones noted were in the lowland forest along the Río Máquina and Río Carizal on April 29, and Aug. 13, 1962.

Laterallus ruber. Red Rail. Very rare; 350 meters; first record. This species was reported by Dickerman to have been heard calling at Arroyo Agrio on the north side of Lago Catemaco between July 22 and Aug. 9, and again on Sept. 3, 1963.

Gallinula chloropus. Common Gallinule. Rare; 0-350 meters; first record. This species was recorded at the north side of Lago Catemaco (1) on March 28, 1960, on the Bahía Sontecomapan, (6) on March 10, and (3) March 29, 1962, and near Punta Roca Partida (1 captured) on March 22, 1962.

Porphyrula martinica. Purple Gallinule. Very rare; 350 meters; first record. Two of these gallinules were feeding at the edge of the marsh at Arroyo Agrio, Lago Catemaco on June 11, 1962. Dickerman reported one there on August 9, and Sept. 3, 1963.

Fulica americana. American Coot. Uncommon; (occasional flocks) 0-350 meters; first record.

Heliorris fulica. Sungrebe. Rather uncommon; 0- meters; first record.

Sun-grebes were noted on the Ríos Máquina and Carizal as well as on Bahía Sontecomapan.

Jacana spinosa. American Jacana. Fairly common; 0-350 meters.

Charadrius collaris. Collared Plover. Status ?; 0 meters; first record. This and the following species were noted occasionally on the Gulf coast.

Charadrius wilsonic. Thick-billed Plover. Status ? 0 meters; first record.

Larus atricilla. Laughing Gull. Fairly common; 0-350 meters.

Gelochelidon nilotica. Gull-billed Tern. Rare; 0-350 meters; first record. This and the following two species were noted at various times on the Gulf coast and over Lago Catemaco.

Sterna albifrons. Least Tern. Rare; 0-350 meters; first record. One was observed over Lago Catemaco on April 10, 1960.

Thalasseus maximus. Royal Tern; 0-350 meters; first record.

Ceryle torquata. Ringed Kingfisher. Fairly common; 0-350 meters.

Chloroceryle amazona. Amazon Kingfisher. Uncommon; 0-350 meters.

Chloroceryle americana. Green Kingfisher. Rather uncommon; 0-350 meters.

Chloroceryle aenea. Pygmy Kingfisher. Very uncommon; 0-450 meters; first record.

Sayornis nigricans. Black Phoebe. Uncommon; 200-500- meters.

Iridoprocne albilinea. Mangrove Swallow. Fairly common; (local); 0-400 meters.

Autobiography

Born in Buffalo, New York, October 28, 1927. Primary and secondary education in Buffalo parochial schools and Saint Joseph's Collegiate Institute. Interest in natural sciences, particularly ornithology, biogeography and meteorology, and participation in field trips and research since early youth. Member of Buffalo Ornithological Society, scientific section of Buffalo Society of Natural Sciences (Buffalo Museum of Science) since 1942. President of section, 1956; elected Fellow, 1959. Since 1949 collected scientific specimens for Buffalo Museum. BA degree, English major, 1948, at Canisius College, Buffalo. Employed by City of Buffalo Health Department, Pest Control Program, 1947-1950. Engaged in field studies in western United States and Mexico, 1950-1952. Military service 1953-1956; intelligence and operations, U.S. Army, Infantry and Corps of Engineers. Duty in Europe included scientific study in six countries, completion of university courses in meteorology, climatology and geography. Joined staff of Buffalo Museum in 1956; appointed Curator, Division of Biogeography, 1959. MA degree, Geography, 1960, at University of Buffalo. Chief instructor, National Science Foundation supported science program at Buffalo Museum, 1959, 1960. Led Museum research expedition to Mexico, 1960. With leave of absence from Museum, commenced graduate work toward doctor's degree at Louisiana State University, 1960, in Departments of Geography and Anthropology, and Zoology. In 1961 awarded grant from National Academy of Sciences to conduct biogeographical research in Mexico. In 1963 returned to Buffalo Museum staff in curatorial position.

www.ingramcontent.com/pod-product-compliance
Lightning Source LLC
Chambersburg PA
CBHW051640170526
45167CB00001B/274